CIRCUITS FOR AUDIO AMPLIFIERS

POWER AMPLIFIERS AND PRE-AMPLIFIERS FOR

MONAURAL AND STEREOPHONIC REPRODUCTION FROM

MICROPHONE, RADIO, TAPE AND PICK-UP SIGNALS

Audio Amateur Press

Publishers
Peterborough, New Hampshire

*This booklet is reprinted by kind permission of
Philips Components Ltd. It was first published by
Mullard Ltd., a wholly owned subsidiary of
Philips Electronics NV, in 1959. Subsequently, Mullard
changed its name to Philips Components Ltd. in 1988.*

*Philips Components Ltd. no longer manufactures the
valves referred to in this publication and regrets it
cannot engage itself in answering any enquiries from
those who build the circuits.*

Second Reprint Edition

Copyright © 1993
Audio Amateur Publications, Inc.
Post Office Box 243
Peterborough, NH 03458-0243

Printed in the United States of America

Reproduction or publication of the content in any manner, without express permission of the
publisher, is prohibited. No liability is assumed with respect to the use of the information herein.

ISBN 1-882580-03-6
Library of Congress Card Catalog Number 93-73046

CONTENTS

CHAPTER		PAGE
	Preface	v
	Mullard Audio Publications	vi
1	Amplifying Systems	1
2	Sources of Distortion in Recorded Sound	4
3	High-quality Amplification	16
4	General Notes on Construction and Assembly	26
5	Twenty-watt Amplifier	29
6	Ten-watt Amplifier	39
7	Three-watt Amplifier	53
8	Seven-watt, D.C./A.C. Amplifier	59
9	Two- and Three-valve Pre-amplifiers	67
10	Four-channel Input-mixing Pre-amplifier	82
11	Three-watt Tape Amplifier	88
12	Tape Pre-amplifier	102
13	Seven-watt Stereophonic Amplifier	113
14	Three-valve Stereophonic Amplifier	121
15	Stereophonic Pre-amplifier	127

PREFACE

The first notable contribution of Mullard design engineers to the specialist world of high-quality sound reproduction was the now-famous 'Five-Ten' amplifier circuit which was introduced in the summer of 1954. In subsequent years, a succession of designs for high-quality circuits has issued from the Mullard Applications Research Laboratory, and these designs include circuits for power amplifiers and pre-amplifiers, circuits suitable for tape-recording equipment and, most recently of all, circuits for stereophonic reproduction.

Many of these circuits have been described in articles in various magazines or have appeared in the form of booklets or leaflets. The others have only recently been released by the laboratory, and have not yet been published. The object of this book is to present the most up-to-date versions of the published circuits together with the new stereophonic circuits in a way which will be useful and convenient to equipment manufacturers, service engineers and home constructors. A list is given on page vi of the earlier Mullard publications which have been incorporated in this book. These publications may be useful for purposes of comparison, but it should be appreciated that the information given here is the latest available and consequently, supersedes that issued elsewhere.

Most of the material contained in this book is based on work undertaken at the Mullard Applications Research Laboratory. The largest contributions have been made by R. S. Babbs, D. H. W. Busby, P. S. Dallosso, C. Hardcastle, J. C. Latham and the late W. A. Ferguson.

While the circuits described in this book have been designed by Mullard engineers, it should be noted that Mullard Limited do not manufacture or market equipment or kits of equipment based on the circuits. Mullard Limited will not consent to the use of the 'Mullard' trademark in relation to equipment based on the circuits described, but will not object to appropriate references to Mullard circuits, specifications or designs. Also, the information contained in this book does not imply any authority or licence for the utilisation of any patented feature.

MULLARD AUDIO PUBLICATIONS

The dates given below are the dates of the first appearance of each publication.
The reference numbers are for the latest reprints of the publications.

5-valve, 10-watt, High Quality Amplifier Circuit	August 1954 (out of print)
High Quality Sound Reproduction*	August 1955 (out of print)
'3-3' Quality Amplifier Circuit	August 1955 (TP352)
Circuits for Tape Recorders	August 1956 (TP430) (out of print)
Supplement to Circuits for Tape Recorders	August 1957 (TP322)
7-watt D.C./A.C. High Quality Amplifier	August 1957 (TP325)
Four-channel Input-mixing Amplifier	August 1957 (TP319)
Two-valve Pre-amplifier	August 1957 (TP339)
Circuit for self-contained Tape Amplifier	August 1958 (TP356)

---

*The section of this book headed 'Design for a 20-Watt High Quality Amplifier' was reprinted from articles which were first published in *Wireless World*, May and June, 1955.

CHAPTER 1
Amplifying Systems

In the last ten to fifteen years, remarkable progress has been made in the field of sound recording and reproduction. The introduction of high-fidelity disc and tape recordings and the improvement in the design of playback equipment have set new standards for the discerning listener and have consequently created the need for systems providing amplification of the highest quality. Because of the variety of signal sources in existence, the range of application of these systems has had to be wide. Therefore, pre-amplifying circuits which cater for pick-up and tape-playback heads, f.m. tuner units and microphones have assumed great importance. Furthermore, the emergence of stereophonic recordings and pick-up heads has created a demand for dual-channel amplification and this has been fulfilled in equipment consisting of either self-contained stereophonic amplifiers or specially-designed stereophonic pre-amplifiers coupled to pairs of conventional power amplifiers. This book contains descriptions and details of high-quality amplifying and pre-amplifying circuits for monaural and stereophonic applications which can be classified as shown in Table 1. Most of these circuits have been designed to be used with each other, and examples of various arrangements of the items of equipment are discussed in subsequent paragraphs of this chapter.

MONAURAL SYSTEMS

The choice of line-up is dictated to a large extent by the strength of the signal to be reproduced. For example, the version of the 10W circuit incorporating volume and tone controls can be used satisfactorily without a pre-amplifier if input signals greater than 500mV are available. (Such signals can be obtained from high-output crystal pick-up heads or f.m. radio tuner units.*) The 3W circuit with volume and tone controls can be used directly with sources which give signal voltages greater than 100mV.

However, signal voltages from low-output crystal pick-ups, magnetic pick-ups and tape-recording heads (which can have values of up to 500mV but are usually between about 2 and 100mV) are normally too low for direct use with the power amplifiers, and a pre-amplifier is necessary to increase the sensitivity of the system. In addition, discs and tapes are deliberately recorded in such a manner that the strength of signals derived from them will be a function of frequency (see Chapter 2). Consequently, the strength of these input signals will usually vary in some way not related

---

* The circuits described in this book can be used with the better-quality f.m. tuner units available commercially. No tuner unit is described here because it is felt that the inherent difficulties of alignment cannot be overcome with the test equipment currently available to the home constructor.

TABLE 1
Classification of Mullard Circuits

Power Amplifiers of General Application
    Five-valve, twenty-watt circuit ('5–20');
    Five-valve, ten-watt circuit† ('5–10');
    Three valve, three-watt circuit† ('3–3').

Pre-amplifiers of General Application
    Three-valve circuit;
    Two-valve circuit;
    Input-mixing circuit.

Power Amplifiers of Special Application
    Seven-watt, d.c./a.c. circuit;
    Three-watt tape circuit (Type A revised);
    Seven-watt stereophonic circuit;
    Two-watt stereophonic circuit.

Pre-amplifiers of Special Application
    Tape-equalising circuit (Type C revised);
    Twin-channel stereophonic circuit.

†Two versions: one with, one without tone and volume controls.

AMPLIFYING SYSTEMS

to the original sound patterns and, to obtain realism from such recordings, some degree of compensation is required in reproducing these sounds. This equalisation is best incorporated in a pre-amplifier, so that the scope of application of power amplifiers can be increased without any resultant fall in the standard

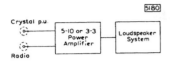

Fig. 1—Direct use of power amplifiers with high-output signal sources

of performance. The 2- and 3-valve pre-amplifier circuits described in Chapter 9 have been designed to precede either the 3W, 10W or 20W amplifiers.

The simple arrangement for the 10W and 3W amplifiers connected directly to the input source is shown in Fig. 1. The use of pre-amplifiers is indicated in Fig. 2.

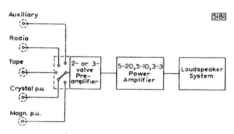

Fig. 2—Use of pre-amplifiers and power amplifiers for low-level signals

The 3W tape circuit comprises a self-contained record-playback system for microphone and radio input sources. The circuit is also suitable for replaying pre-recorded tapes. A programme-monitoring signal is available when recording and, during playback, this source will provide a low-level signal suitable for driving a power amplifier if an output power higher than 3W is required.

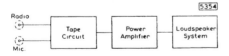

Fig. 3—Tape circuits and power amplifiers arranged for use with radio or microphone signals

The tape pre-amplifier provides recording facilities comparable with the self-contained unit but, on playback, the circuit only provides an equalised signal for driving a power amplifier. A simple arrangement for the combination of tape pre-amplifier and power amplifier is shown in Fig. 3. This line-up is suitable also for the 3W tape amplifier when the low-level playback output is used.

For a complete record/monitor-playback system using pick-up input sources, the tape pre-amplifier should be used in conjunction with the line-up of Fig. 2. The new arrangement is shown in Fig. 4*. When recording, the input signal is fed to the appropriate input channel of the pre-amplifier and then from the programme-recording output of the pre-amplifier to the input of the tape circuit. The programme can

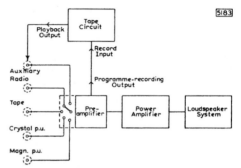

Fig. 4—Arrangement of pre-amplifiers and power amplifiers with complete tape record/monitor-playback facilities

be monitored by using the normal output of the pre-amplifier to drive the power amplifier. For playback, the equalised output from the tape pre-amplifier is fed via the Auxiliary input channel to the main amplifying chain.

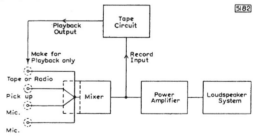

Fig. 5—System for programme recording, monitoring and replaying with provision for mixing input signals

To feed several inputs into an amplifying system at the same time, the main pre-amplifier in Fig. 2 should be replaced by the input-mixing pre-amplifier. Recording, monitoring and playback are possible with the new arrangement shown in Fig. 5.* The output impedance of the mixer is low so that the inputs of the tape pre-amplifier and the power

---

* It is important that the tape deck in both these arrangements should always be switched to the playback position before the connection from the tape output to the pre-amplifier input is made. If the tape recorder is in the record position when the connection is made, oscillation will occur.

amplifier can be connected simultaneously to the mixer without affecting the quality of reproduction.

STEREOPHONIC SYSTEMS

For stereophonic reproduction, two separate amplifying channels are required, one for each of the twin signals obtained from the stereophonic signal source. In the simplest arrangement, the 2W or 7W stereophonic amplifier is connected to the source as shown in Fig. 6. If the stereophonic pre-amplifier is used, it should be coupled to two power amplifiers and loudspeaker systems as in Fig. 7.

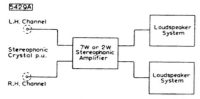

Fig. 6—Use of stereophonic amplifiers

Stereophonic systems should be suitable for reproduction from monaural sources, and 'compatibility' switches are usually incorporated for this purpose. For instance, the switching indicated in the pre-amplifier stage of Fig. 7 selects the input channels* and also sets the equipment for normal or reversed stereophonic reproduction or for monaural reproduction using both power amplifiers and loudspeaker systems.

POWER SUPPLIES

Facilities are provided in the 10W and 20W amplifier circuits for obtaining power supplies for a single-channel pre-amplifier or f.m. radio tuner unit. For stereophonic applications, the h.t. supply for each channel of the pre-amplifier can be taken from the power amplifier used with that channel.

The h.t. supply facilities for pre-amplifiers or
___
* Facilities are provided for stereophonic radio signals in anticipation of stereophonic transmissions.

tuner units are not immediately available in the 3W circuit, but the modifications required are not extensive. Sufficient current is available with the power-supply stages shown in Chapter 7 for either the 2- or 3-valve pre-amplifier. The h.t. supply to the pre-amplifier can be taken from the reservoir capacitor of the supply circuit but it will need extra smoothing. The components required for this are indicated in the chapter on pre-amplifiers (Chapter 9), but they should be mounted on the chassis of the main amplifier. The current drain when an f.m. tuner is coupled to the amplifier is increased considerably and it will be

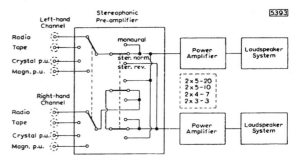

Fig. 7—Arrangement of stereophonic pre-amplifier with pairs of conventional power amplifiers ('4–7' power amplifier refers to 7W d.c./a.c. circuit)

necessary to choose a rectifier and mains transformer which will supply this current. Again, extra stages of decoupling will be required to prevent excessive hum voltage being picked up in the sensitive early stages of the tuner.

A separate power stage is described for the tape pre-amplifier, and this can be retained when the pre-amplifier is used with any of the power amplifiers. Physically, however, the extra unit may prove an encumbrance and consequently it may be desirable to modify (in the manner suggested above) the power stage of the main amplifier to supply the needs of the tape pre-amplifier.

NOTE: The word *monaural* is at present commonly used to depict conventional sound reproduction in which a single-track recording and a single pick-up head (or tape playback head) are used. The word *monophonic* may often be encountered as a synonym of *monaural* and may eventually supersede it. *Monaural* means, literally, *of one ear* and a *monaural* effect is that of listening with one ear, when no impressions of space or displacement are conveyed to the listener. *Monophonic* means *of one sound source* so that *monophonic* reproductions again lack any impressions of spread or position. *Monaural* has been adopted in this book simply because it is currently more popular: its use should not be taken to imply any etymological superiority of the word over *monophonic*.

CHAPTER 2
Sources of Distortion in Recorded Sound

Some distortion is always present in the sound reproduced from a gramophone record or magnetic tape. It is inevitable that part of this distortion will be distortion recorded on the disc or tape itself, but the loudspeaker, pick-up and recording head also make substantial contributions. A carefully constructed amplifier is far from being the weakest link in the chain: in fact, it is improbable that the limit to the quality of reproduction which can be obtained will be set by the performance of the amplifier. High-quality reproduction will only be achieved if careful attention is also paid to the sound input source and to the loudspeaker.

Even if *perfect* associated equipment could be obtained, however, it would not be possible to eliminate distortion entirely. There would still be some distortion – a comparatively small amount it is true – which is bound to occur when discs or tape are used. The function of this chapter is to give brief accounts of the main items of associated equipment needed for disc or tape reproduction and also to discuss some of the causes of distortion, indicating the methods by which imperfections are minimised.

LOUDSPEAKERS

The choice of a suitable loudspeaker is very important for both disc and tape reproduction. While it would be wrong to assume that price is the only guide available, normally the higher grades of loudspeaker will justify their extra cost by giving correspondingly better reproduction. The choice should never be made hastily. It is not sufficient to listen to one or two records because it is very easy to construct a system which will sound impressive at first hearing. Extended periods of listening to music are desirable to discover if mental fatigue is likely to be caused by the system. Furthermore, because it is easier to judge speaker quality by listening to comparatively simple sounds rather than to full orchestral recordings, the system should also be tested by assessing the fidelity with which good-quality broadcast speech is reproduced.

A simple, direct-radiator loudspeaker – that is, a single-cone unit handling the entire range of frequencies – can give excellent results when mounted in a suitable enclosure. It is important to remember that this enclosure should always be that specified by the speaker manufacturer since it is not possible to design a suitable cabinet without knowing accurately the basic performance of the loudspeaker unit. One of the main disadvantages of all types of simple direct radiator is that the high-frequency response is very directional. This fact, however, can to some extent be overcome by using the loudspeaker in conjunction with some form of reflector. The reflector can be built into the cabinet, or the corner of a room can be used for the purpose. The effect of the reflector is to distribute the high-frequency components in a more random manner so that, with the adequate treble boost provided by the amplifier, good results will be obtained.

There is no doubt, however, that even the simplest of dual systems is much better than a single unit. Such a dual system may consist of two direct radiators – for example, a 12 in. unit for the bass-end of the spectrum and a 5 in. unit for the middle and treble components – used in conjunction with a suitably designed crossover network. It must be remembered that, while the effective frequency response may not be very much increased by this type of dual system, various forms of distortion will be reduced considerably, added to which there will be a better distribution of high frequencies from the smaller source.

Dual systems using twin direct radiators are available in two forms: either with two completely separate units, each mounted in its own enclosure or, alternatively, with the high-frequency unit mounted on some form of modified frame, making it an integral part of the low-frequency unit.

LOUDSPEAKERS

Greater improvement may be obtained by using three loudspeaker units together with a three-way cross-over network. This extends the frequency range and reduces the distortion still further, as well as providing a wider range of distribution patterns.

Normally, suitable cross-over networks are recommended by the loudspeaker manufacturers, and it is usually best to adopt these recommendations. However, an example of a two-way network having a very compact design and unusually low losses is shown in Fig. 1. Two Mullard Ferroxcube pot-core inductors, type LA23, are used in conjunction with the special metallised-paper capacitors now available, and a maximum rate of attenuation of 12dB per octave can be obtained from the constant-resistance circuit.

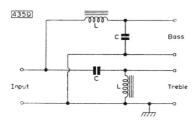

Fig. 1—Two-way loudspeaker cross-over network

The values of L and C in the diagram are given in terms of the loudspeaker impedance R and the cross-over frequency f by the following equations:

$$L = \frac{R}{\sqrt{2}\pi f},$$

$$C = \frac{1}{2\sqrt{2}\pi f R}.$$

The approximate number of turns required on the inductor for a given inductance L is $35\sqrt{L}$, where L is in millihenries.

In practice, the impedance R of the speaker will already be determined, and it will be convenient to obtain the value of capacitance to give the required cross-over frequency from the appropriate section of Table 1. The corresponding value of inductance, and the approximate number of turns to give this value, can then also be obtained from the table.

One advantage of horn-loaded loudspeakers over direct-radiator types is that the effective damping is increased, and, consequently, there is improved reproduction of transients. Some units incorporate a combination of a direct-radiator bass speaker and a horn-loaded treble speaker. They enable a higher conversion efficiency to be obtained in the treble region of the spectrum, thus ensuring a more even distribution for a given frequency response. In other words, the total energy output of the high-frequency unit is maintained up to a higher point in the musical scale. Although it is possible to use a completely separate horn-loaded tweeter, at least two manufacturers use a double-magnet system which permits the treble unit to be mounted concentrically with the bass unit. Thus, phase distortion at the cross-over frequency, which is produced in other systems by the distance between the two units, is also avoided. Also, the overall frequency response can be made smoother with the concentric system.

TABLE 1

Component Values for Cross-over Network (Fig. 1)

Loudspeaker impedance R (Ω)	Cross-over frequency f (c/s)	C (μF)	L (mH)	No. of turns	Wire gauge (s.w.g.)
15	470	16	7·1	93	20
	620	12	5·4	81	20
	750	10	4·5	74	20
	940	8	3·55	66	20
	1250	6	2·7	57	20
	1880	4	1·8	47	18
	3750	2	0·9	33	18
7·5	470	32	3·55	66	20
	620	24	2·7	57	20
	940	16	1·8	47	18
	1880	8	0·9	33	18
3·75	940	32	0·9	33	18
	1250	24	0·67	29	18
	1880	16	0·45	23	18

The most complicated form of dual loudspeaker system is that which is horn-loaded over its entire range. This may be achieved by using separate horn assemblies, each associated with a loudspeaker unit covering part of the frequency spectrum. Alternatively, loudspeakers of the concentric type can be used. Such systems are always expensive but they have certain advantages: the electro-acoustic conversion efficiency is high, the acoustic damping for the entire audio range is much improved over that of any simple direct-radiator system, and the larger size of bass horn gives better realism.

DISC EQUIPMENT

RECORDING CHARACTERISTICS

Cutting Styli

In making records on discs, the recording head is used to convert the signal voltage (or voltages, in stereophony) from the recording amplifier into vibrations of the cutting stylus. The stylus cuts the groove in the disc during recording, and the vibrations are reproduced in this groove as programme modulation. With monaural recordings, the single programme modulation appears as an identical* trace in both walls of the groove. With stereophonic recordings, however, the twin signals give rise to a more complex trace in which each wall of the groove carries one signal.

The cutting edge of a monaural stylus is rectangular and it cuts with a lateral movement so that the width of the recording groove is constant*. A stereophonic cutting stylus is triangular and has two cutting edges which are set symmetrically about the vertical at right-angles to each other. Each signal causes a cutting edge to move in a direction perpendicular to its length, so that the resultant movement of the stylus is not a simple one from side to side as with a monaural stylus, but is two-dimensional in the plane of cutting edges. Thus, whereas a monaural stylus needs only freedom of lateral movement, a stereophonic stylus must have vertical compliance as well.

Characteristics

Except for the duality of stereophonic discs, the characteristics of monaural and stereophonic recordings are fundamentally the same. Therefore, it is sufficient in these simple comments on recording characteristics to refer only to monaural recordings. The remarks can be applied, with suitable allowance for their two-fold properties, to stereophonic records.

Magnetic recording heads are normally used for cutting discs: the stylus is attached either to the moving iron armature or to the moving coil. The velocity at which the stylus vibrates (measured as the stylus passes its equilibrium position) is directly proportional to the recording-signal strength, so that, for a constant voltage, the velocity is constant, and the amplitude or width of the groove is inversely proportional to the signal frequency. This is known as 'constant-velocity' recording.

If the amplitude at high frequencies is chosen to give an acceptable signal-to-noise ratio, the amplitude at low frequencies will be excessive. Distortion will be high and too much space will be needed between adjacent rings of the groove to ensure that breakthrough from one ring to another does not occur. Thus, some restriction of the amplitude is desirable at these low frequencies and, to achieve this, bass signal voltages from the recording amplifier are attenuated before being recorded. Recordings in which the maximum width of the groove is restricted to some limit are known as 'constant-amplitude' recordings.

Because of the difficulty in maintaining a satisfactory signal-to-noise ratio if a constant-velocity characteristic is used at high frequencies, it is normal to boost treble signals from the recording amplifier before recording them. Consequently, a recording characteristic indicating the variation of recording voltage with frequency can be divided into three sections: the bass section, showing voltage attenuation; the middle section, showing constant-velocity recording; the treble section, showing voltage boost.

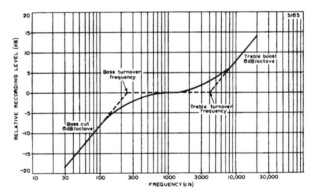

Fig. 2—Hypothetical frequency characteristic used for recording on discs. The dotted line shows constant-amplitude and constant-velocity regions; the full line shows practical realization of these.

An imaginary recording characteristic is drawn in Fig. 2. The frequencies at which the characteristic changes from one section to another are called the 'turnover' frequencies. The slope of the bass and treble sections of the characteristic depends on the degree of attenuation or boost applied to the recording signal. If recordings with a constant amplitude of modulation are required, the stylus velocity must be proportional to the frequency and hence the signal voltage must also be proportional to the frequency. Thus, the signal voltage must be halved or doubled as the frequency is halved or doubled and this gives a characteristic slope of 6dB per octave.

---

*See "Tracing Distortion", page 9.

DISTORTION FROM DISC EQUIPMENT

Actual recording characteristics differ considerably from the imaginary one of Fig. 2. The straight lines of the characteristic cannot be achieved in practice, and actual characteristics are more like the continuous curve shown in Fig. 2.

ation should be the converse of the attenuation and boost applied while recording.

The recording characteristics used by the different recording companies before 1955 followed, to a greater or lesser extent, the curve shown in Fig. 2, but the

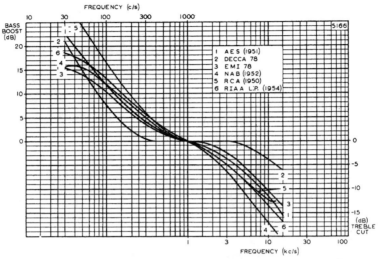

Fig. 3—Commercial playback characteristics for disc equipment

The bass and treble turnover frequencies in actual curves are defined as those frequencies at which the ratios of the recorded voltages to the true signal voltage are $1:\sqrt{2}$ and $\sqrt{2}:1$ (that is, -3dB and $+3$dB) respectively. The true signal voltage is assumed, for reference purposes, to be the level of voltage recorded at 1kc/s.

In most practical characteristics, the middle horizontal section is discarded and only a point of inflection at 1kc/s retained. The bass and treble sections of practical curves are not produced indefinitely with a constant slope, but are flattened at the limits of the range of audible frequencies.

PLAYBACK CHARACTERISTICS

If a magnetic pick-up head is used to reproduce the sound recorded on discs, the output voltage from the pick-up will be proportional to the velocity of vibration of the needle. Consequently, the output from a magnetic pick-up used with recordings made to characteristics of the type shown in Fig. 2 will increase with frequency in the bass and treble regions. (This will not be so with crystal pick-ups because the output of these is proportional to the amplitude of modulation.) Thus, correction or 'equalisation' will be required during playback amplification to restore the true level of the signal voltage, and this equalis-

differences were large enough to make equalisation a matter depending greatly on the records to be played. Many companies issued their own equalisation characteristics, and corrective networks had to be designed with these in mind. Examples of the playback characteristics of the major recording

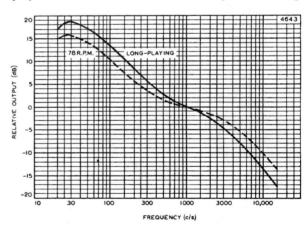

Fig. 4—RIAA playback characteristics for standard and microgroove discs

companies are shown in Fig. 3. In 1955, however, the majority of these companies agreed to adopt the characteristics of the Recording Industries' Association of America (RIAA), which are drawn in Fig. 4.

DISTORTION FROM DISC EQUIPMENT

Consequently, there is no need, with up-to-date recordings, to adapt the pick-up head and record to each other by using separate equalising networks for each brand of record. Instead, one network (usually in the form of a frequency-selective feedback network) can be incorporated in the amplifier. Of course, with the two characteristics for standard and microgroove records, separate networks will still be required for these.

PICK-UP HEADS

The principle on which a gramophone pick-up head operates can be one of several, but the commonest types of component belong to either the constant-velocity or the constant-amplitude group. Magnetic pick-up heads (moving-iron and moving-coil) have basically constant-velocity characteristics, the output voltage being proportional to the velocity of vibration of the pick-up stylus. Piezo-electric (crystal) heads possess constant-amplitude characteristics, the output from these varying with the amplitude of modulation in the recording groove.

As the RIAA recording characteristics (and, similarly, most of the commercial characteristics before 1955) do not possess purely constant-velocity or constant-amplitude properties, some degree of equalisation will be needed with either group of pick-up head, and the degree will be different for each group. However, simple resistance-loading of low- or medium-output crystal heads will convert their characteristics into very near approximations of constant-velocity characteristics without prejudicing the standards demanded of high-quality equipment. Consequently, this method of loading is often adopted in high-quality amplifiers to give a simpler switching arrangement of the equalising feedback networks. With it, the networks provided for standard and microgroove records using magnetic heads can also be used for medium- or low-output crystal heads.

Magnetic Pick-up Heads

Both types of magnetic pick-up head – that is, moving-iron and moving-coil – work on the principle of the dynamo: the change in magnetic flux through a conductor generates a voltage across the conductor. The change of flux in the heads results from the movement of either the magnetic core or the conducting coil, and this movement is caused by the vibration of the stylus (attached to either the core or the coil) in the recording groove.

The output voltage from either type of head is proportional to the rate of change of flux through the conductor and thus to the velocity of the stylus (measured at its equilibrium position). Output voltages which are quoted for comparing magnetic pick-up heads should therefore be given with reference to the stylus velocity. This reference can be made directly by quoting the velocity; alternatively the ratio (expressed in decibels) of the actual output voltage to that which would be obtained at some standard velocity (1cm/sec) should be given. Hence, an output voltage could be given as, say, 100mV at 3·16cm/sec or 100mV at +10dB, and both expressions would be equivalent.

Moving-iron magnetic pick-up heads can be divided into three classes, governed by their outputs (measured at a stylus velocity of 3·16cm/sec):

Low-output heads:	Voltages below 20mV
Medium-output heads:	Voltages between 20 and 100mV
High-output heads:	Voltages above 100mV.

The basic output of a moving-coil pick-up head is of the order of millivolts. The lower level of output results from the desirability of lightness of the pick-up unit and hence the restriction on the number of turns of the coil.

Both types of head are suitable for high-quality equipment, although the intrinsically low output of the moving-coil type necessitates the use of a high-ratio pick-up transformer between it and the amplifier. The impedance of magnetic heads is of the order of kilohms at signal frequency of 1kc/s, and the required loading (that is, the input impedance of the amplifier) for the rated output of the heads is of the order of 100kΩ. Their loading should be independent of frequency, otherwise the input signal to the amplifier will also vary with frequency.

Crystal Pick-up Heads

A voltage is developed between the faces of a piezo-electric crystal when the crystal is strained. The voltage is proportional to the strain. In a crystal pick-up head, the playback stylus is attached rigidly to some piezo-electric material, and a strain is produced in the material by the vibration of the stylus as it follows the modulation of the recording groove. The amplitude of this modulation governs the output voltage of the head.

The rated output voltage of a crystal pick-up head is always quoted at a particular frequency. This is essential because, for comparison purposes, a recording made to a constant-velocity characteristic is used and the amplitude of modulation – and, therefore, the output voltage – of such a recording is inversely proportional to the frequency. A typical output voltage from a high-output crystal head would be about 1V at 1kc/s. The output at a frequency of 300c/s would therefore be about 3V. It is thus evident that attenuation is desirable at the lower frequencies to provide a balanced output and to prevent the input valve from being over-driven. The highest-quality crystal heads have outputs considerably lower than this, a typical value being about 0·5V at 1kc/s.

The impedance of crystal pick-up heads is equivalent to a capacitance, the value of which is of the order of 1000pF. The optimum load resistance of a crystal head will depend on the method of loading. If it is loaded so that its characteristic is similar to that of a magnetic pick-up then the load resistance is of the order of 100kΩ. If the characteristic is not modified, then an input impedance of the order of megohms is desirable.

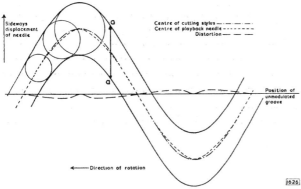

Fig. 5—Illustration of tracing distortion (Sine-wave modulation of recording groove is greatly magnified)

Stereophonic Pick-up Heads

Pick-up heads for use with stereophonic equipment can be of either the magnetic (moving-iron) or the crystal type. The construction of these pick-up heads, however, will differ considerably from that of monaural heads. They must be capable of responding to the two-dimensional modulation of the recording groove and of translating the response into two separate signals, whereas monaural heads have only to follow a lateral displacement and convert it into a single signal. Consequently, there must be two signal generators in each stereophonic head and these must have very similar characteristics. They will therefore both be of the same type – both crystal or both magnetic. Also, the pick-up stylus will require a vertical compliance comparable with its lateral compliance so that the tip of the stylus can follow the contours of the recording groove faithfully and provide balanced stimuli for the twin generators.

The output voltages obtained from these generators are very much lower than the voltages derived from monaural heads of the same type. Moving-iron, stereophonic heads will produce voltages of the order of millivolts for each signal and crystal heads will give signals of the order of 100mV. The impedance of each section of a stereophonic head will be of the same order as that of the same type of monaural heads – about 1 kilohm for magnetic heads and the equivalent of about 1000pF for crystal heads.

DISCS

Records are normally made for replaying at three different speeds: 78, 45 or 33⅓ revs/min. The first or 'standard' type of disc is pressed in a hard non-pliable material called shellac. Record wear with this material is relatively high even under good playing conditions. Wear with vinyl, the pliable material used for the 45 and 33⅓ revs/min, 'microgroove' type of disc, is very slight under comparable conditions. The main trouble with vinyl records is the tendency to collect dust through electrostatic attraction.

Standard records are made with diameters of 10 or 12 inches and provide playing times of about 4 or 5 minutes. 'Extended-play' or 45 revs/min discs have diameters of 7 inches and playing times of about 10 minutes. 'Long-playing' or 33⅓ revs/min records are made with diameters of 10 and 12 inches and play for up to about 30 minutes. Stereophonic discs are produced in 'Extended-play' and 'Long-playing' forms.

Several types of pick-up stylus are available (particulary for use with standard monaural records), but the most common for both standard and microgroove discs is the permanent or jewel-tipped stylus. Diamond styli are the hardest wearing, but sapphire types are very popular.

TRACING DISTORTION

One form of distortion which occurs with monaural disc recordings is that resulting from the difference in shape of the recording groove and the rounded playback stylus. The way this distortion arises can best be seen by considering the width of the recording groove.

When no sound is recorded, the groove is unmodulated and the cutting face of the stylus is at right-angles to the length of the groove. The width of the groove is uniform and is the full width of the cutting edge. With a signal, however, the stylus will be displaced from the unmodulated position and the cutting face is at an acute angle to the direction of the groove. Thus the width of the groove is not uniform. This is illustrated in Fig. 5, which shows a sine-wave groove, very much magnified.

DISTORTION FROM DISC EQUIPMENT

The width of the groove in the direction *a-a* always corresponds to the full width of the cutting edge of the stylus. The true width of the groove – that is, the width at right angles to its length – therefore depends on the angle between the modulated groove and the direction of the unmodulated groove, and will be smaller for greater values of this angle. The true width of the groove is the same as the width of the cutting edge only at the peaks of the sine wave.

If distortion is to be avoided, the sideways movement of the playback stylus should reproduce exactly the movement of the cutting stylus. The circles in Fig. 5 represent the point of the pick-up stylus as it rests in the groove during playback, and the centres of these circles should therefore lie on the line traced by the centre of the cutting stylus. The figure shows that this condition is not fulfilled: the dotted line traced by the centre of the pick-up stylus is not the same as the chain line traced by the centre of the cutting edge. The amount of distortion resulting from this lack of coincidence depends on the difference between these two lines, measured parallel to *a-a*. There is no distortion at the peaks of the sine wave, nor in the unmodulated position.

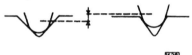

Fig. 6—Illustration of 'pinch effect'

Tracing distortion consists of odd harmonics. The distortion increases as the modulation becomes more spiky. Thus, it increases with the amplitude of modulation – one reason why constant-amplitude recording is necessary at low frequencies. It also increases with frequency, which is a reason for restricting the amount of treble boost used when recording. The distortion is also greater if the modulation is compressed, as it will be near the centre of the record.

PINCH EFFECT

The sine-wave groove of Fig. 5 illustrates the so-called 'pinch effect' which results from the varying width of a modulated groove on a monaural recording. Because of this varying width, the pick-up stylus will have to ride up and down as it traverses the groove (Fig. 6). At the end of a microgroove recording, the maximum vertical movement may be as much as 11% of the maximum lateral displacement. The output of some pick-up heads varies slightly with this vertical movement so that the pinch effect can be an additional source of distortion. As the needle has to move up and down twice during each cycle, the distortion is of even harmonic order.

NEEDLE-SCRATCH

As the pick-up stylus traces the groove, the fine particles in the material of the disc cause the stylus to make small irregular movements. The voltage set up by these movements is amplified and is heard as needle-scratch. The noise lies mainly between 2 and 10kc/s.

Needle-scratch will be particularly bad when the diameter of the needle is too small for the groove (Fig. 6). For good reproduction the stylus should rest on the two sides of the groove. If it rests right in the trough, it may ride up the walls, thus giving excessive scratch and additional distortion. If it rests on top of the groove only, it cannot follow the groove correctly, and there will then be a tendency for the stylus to 'skate' across the record.

TRACKING ERROR

Another important difference between the making and replaying of discs is that the pick-up stylus as it traverses the record does not follow the same path as the cutting stylus. The cutting stylus is set to move inwards in a straight line along a radius of the disc, whereas the pick-up stylus must necessarily be mounted on a pivoted tone arm, and must cross the record in an arc of a circle. The playback stylus is thus not always held at right-angles to the groove and the output will vary slightly. This effect is greater for large records, and can be reduced by using a longer pick-up arm.

TURNTABLE DRIVING SYSTEMS

It is essential that the mechanical driving systems of disc-playing equipment should be accurately made, especially with stereophonic equipment. Rotation of the turntable must be extremely steady. These requirements must necessarily be very severe if the mechanical system is to match the performance expected from the acoustical and electronic equipment.

In a high-quality installation in which the loudspeaker is separate from the main chassis, the system can be mounted solidly to the motor board, and the motor board itself floated on rubber to ensure that no mechanical vibration is transmitted to the cabinet.

An unsuitable system can spoil the loudspeaker output in three objectionable ways: 'wow', 'flutter' and 'rumble'.

Wow and flutter are produced by variations in the speed of the turntable. Those variations occurring at low frequencies are termed wow; flutter results from faster fluctuations. Slow variations can be caused by spindles and other rotating parts being loosely mounted or off-centre, so that the wow resulting is

DISTORTION FROM TAPE EQUIPMENT

heard at regular intervals. Unevenness of the drive, changes of friction (catching or slipping) and record-slip can give rise to intermittent wow. Small irregularities are normally experienced when driving from a motor. Sudden changes in speed of as little as 0·5% will be clearly noticable as flutter when a long note is being played. A sufficiently steady speed is usually achieved by providing a heavy turntable, or some other moving part, to act as a flywheel. Care should be taken to see that the speed-changing mechanism is not subject to undue wear. This can lead to wow or flutter after much use.

Rumble results because the mechanical parts run noisily. These parts should run sufficiently quietly for no objectionable rumbling to be heard at full gain when there is no other input signal.

TAPE EQUIPMENT

RECORDING AND PLAYBACK HEADS

Generally speaking, tape decks (with the exception of some of those used with the so-called 'professional' equipment) combine the duties of recording and reproduction in a common head. The requirements of the recording and the playback heads are fundamentally the same and, as they are not required simultaneously, economy is served by combining them.

Modern heads are usually constructed of two semi-circular stacks of high-permeability laminations of about a half to one inch in diameter. The stacks are assembled symmetrically with a 'head gap' (against which the tape passes) of about five ten-thousandths of an inch, and an auxiliary gap, diametrically opposite the head gap, of about ten times that width. Both gaps are filled with non-magnetic, metallic 'shims' to prevent any accumulation in the gaps of magnetic material which may be rubbed off the tape by friction, and to maintain a linear relationship between the flux at the gaps and the current through the exciting coils of the head.

The impedance of these coils, which are placed symmetrically about each core, should, for recording, be low compared with the source impedance. This ensures, as far as is possible, a current which is independent of the frequency of the signal. For the playback process, however, maximum output voltages are to be gained from high impedance windings. Any conflict between these requirements can be resolved by feeding the output from low impedance coils into a step-up transformer.

Stereophonic playback heads consist fundamentally of two monaural heads mounted one on top of the other. Each head forms part of a complete playback system, the input signal for which is derived from one of the tracks of specially recorded stereophonic tape. Facilities for domestic stereophonic recording are seldom provided because of the difficulties to be encountered and the advanced technique required for such an operation.

ERASE HEADS

Although similarities exist between the erase head and the recording head, they differ in construction in two principal respects: Firstly, appreciable power (about 2 to 4W) is required for efficient erasure, while that necessary for the recording head is usually about 1mW. Hence the core of the erase head must be of a material having a higher point of magnetic saturation than the material in the recording head. Secondly, the longer the tape is under the influence of the erasing field, the more complete will be the cleaning. Thus the gap in the erase-head core should be considerably larger than in the core of the combined head. The erase gap is usually about fifteen thousandths of an inch wide. Normally there is no auxiliary gap in the core of the erase head.

TAPES

Modern magnetic tapes consist, generally, of a non-magnetic base (paper, cellulose acetate or p.v.c., for example) coated with a magnetic material. The base obviously needs to have strength and suppleness. The magnetic material (often red or black iron oxides) is a very finely divided powder mixed in some binding substance (lacquer, for instance), and the coating applied to the base material has to be very smooth. Unevenness in the surface of the tape will tend to cause broken contact with the heads, and consequently an undulating level of recording. The magnetic material needs a high coercivity to prevent, in particular, very large demagnetisation losses at treble frequencies, and a high remanence to give a good level of recording.

Tapes are usually supplied in 5- or 7-inch reels, giving playing times, depending on the transport speed and the thickness of the tape, of between about 10 and 90 minutes. The standard width of the tape is a quarter of an inch, which normally allows for two adjacent recording paths. With stereophonic tapes, the adjacent paths are used for the twin signals comprising the programme. The total playing time of such tapes is therefore only half that of monaural tapes.

DISTORTION FROM TAPE EQUIPMENT

TAPE TRANSPORT SYSTEMS

The basic requirement of the transport system is that a steady speed should be imparted to the tape. Momentary changes in this speed will cause 'wow' or 'flutter'. (Wow results from slow variations; flutter results from fast fluctuations: the dividing line is arbitrary.)

The actual speed of transport has a considerable bearing on the performance of the recording apparatus. The higher the speed, the better the performance at the higher frequencies. This is closely allied to the size of the gap in the recording head. But, of course, economically, great speeds are a disadvantage: the playing time of the tape is obviously reduced. Most transport systems are standardised for playing speeds of either $3\frac{3}{4}$, $7\frac{1}{2}$ or 15 inches of tape per second, with possibly, a choice of speeds.

AMPLITUDE DISTORTION

The non-linear relationship between the magnetism residing on the tape and the magnetic field inducing it, will cause considerable amplitude distortion in the recording.

The curve OB_3B_5 in Fig. 7 shows a typical relationship between the magnetisation produced in a magnetic material and the increasing magnetic field producing it. A field strength of H_2 units, for example, would produce a degree of magnetisation corresponding to the point B_2, provided the material was unmagnetised initially, and also provided the field strength was not, at any time during its general increase from zero to H_2, caused to diminish. The levels of magnetisation corresponding to B_1, B_2, etc., are those obtaining when the field strengths are actually H_1, H_2, etc. These do not represent residual magnetism. If the fields are reduced from H_1, H_2, etc., to zero, the induced magnetism decreases along the paths B_1R_1, B_2R_2, etc., respectively, and the points R_1, R_2, etc., denote the degrees of residual magnetisation, or remanence, induced in the magnetic material by maximum applied field strengths of H_1, H_2, etc.

The curve OC_3C_5 plotted in Fig. 7 gives a typical 'transfer' characteristic for a magnetic material; that is, it shows the intensity of residual magnetism (R_1, R_2, etc.) resulting from any given maximum magnetising field (H_1, H_2, etc.). Complete transfer characteristics for positive and negative magnetising fields are shown in both Figs. 8 and 9.

It is obvious from these characteristic curves that the degree of residual magnetism is not a linear function of the maximum magnetising field. Because of this non-linearity, any signal applied to the microphone would, on recording, suffer distortion. All signals (for example, the 'unbiased' sine wave shown in Fig. 8) would suffer 'bottom-bend' distortion resulting in even harmonics of the fundamental appearing in the recording. Large signals would be further distorted by the magnetic saturation depicted by the flattening of upper sections of the characteristics, and the 'clipped' reproduction would contain a high percentage of odd harmonics of the original.

The middle sections of the arms of the transfer characteristics are, however, approximately linear, and if the variation in the magnetic field strengths are confined to these portions, then the recording will

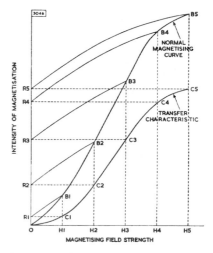

Fig. 7—Variations of instantaneous and residual intensities of induced magnetism with the strength of magnetising field

be a relatively undistorted replica of the original sound. If a constant current is fed into the recording head together with the signal current, the effect is to 'lift' the variations in the magnetising field above the lower curvature of the characteristic (Fig. 8). The d.c. bias induces a constant degree of magnetisation in the tape on which the audio variations are superimposed. Only the variations of the residual magnetism will appear on replaying the recording, so that the bias will not be translated into sound (in fact, some noise does result from the biasing remanence). But because of the limitation imposed by the saturation curvature of the characteristic, the fact that the bias magnetism is added to the audio magnetism means that the amplitude of the signal must be restricted quite considerably. The signal-to-noise ratio is thus lowered when d.c. bias is used.

This method of preventing amplitude distortion by introducing a constant biasing current into the recording head together with the signal current has

DISTORTION FROM TAPE EQUIPMENT

been superseded by a method in which the direct current is replaced by a high-frequency, alternating current. The result is a recording relatively free from harmonic distortion in which a good signal-to-noise ratio is maintained. No complete explanation of

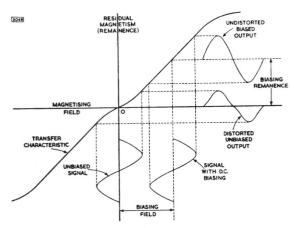

Fig. 8—Distortion occurring with no bias, and its removal with d.c. biasing

the mechanism of a.c. biasing has yet been accepted generally, but some idea of the process can be obtained from Fig. 9.

A feature of this method of biasing is the fact that no residual magnetism is induced in the tape by the

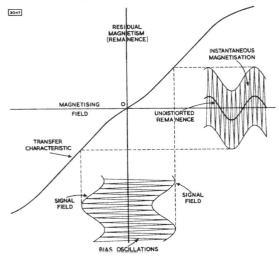

Fig. 9—Removal of distortion with a.c. biasing

a.c. signal, provided, of course, that these oscillations are free from even harmonic distortion. As any point on the tape passes the gap in the recording head, it is subjected to a rapidly alternating magnetic field, the strength of which increases as the point approaches the gap and dies away as the point recedes. Such a

process causes no remanence in the tape. Only the audio variations, which are superimposed on the a.c. field, cause any residual magnetism, and, because of the bias, this magnetism depends linearly on the audio signal. Thus, the tape is not loaded with any 'wasted' remanence as it is when d.c. biasing is used. In addition to this, it will be seen from Fig. 9 that both linear sections of the transfer characteristic can be used. It follows then that the limitations imposed by d.c. biasing on the strength of the audio signal can be relaxed considerably without increasing the danger of approaching the saturation level of the tape. A.C. bias, therefore, reduces 'bottom bend' distortion and, compared with d.c. biasing, gives an improved signal-to-noise ratio.

Initially, an increase in the a.c. biasing current for a given signal strength lessens the distortion of the

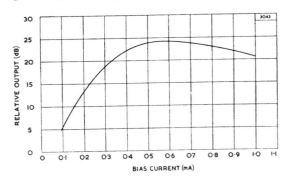

Fig. 10—Variation of output with bias current

recorded signal and increases the level of recording. But too much bias will cause 'top-bend' distortion and will also cause a lessening in the output, especially for treble signals. (The bias acts very much in the manner of the erase signal – see below.) A typical variation of output with bias current is shown in Fig. 10.

ERASING

It was stated above that the field associated with the a.c. biasing signal did not cause any residual magnetism in a tape passing through the field. If the tape is unmagnetised when the signal is applied, it will be unmagnetised when it has passed through the field. If the amplitude of the a.c. signal is sufficient to saturate the tape, then even if the tape is not unmagnetised originally, it will be so when it has travelled through the field.

This fact is used in the erase heads of most magnetic recording equipment. The tape is drawn through the strong local magnetic field arising from a high-frequency alternating current in the erase head. As each point of the tape travels into the field, it is

DISTORTION FROM TAPE EQUIPMENT

saturated, and as each point recedes, it is subjected to a field strength which diminishes to zero some distance from the head, By this process it is cleaned of all previous magnetisation.

OSCILLATOR COIL

The bias oscillator should generate a sine wave with negligible harmonic distortion to keep noise to a minimum. Symmetry in the waveform is essential. An asymmetrical sine wave has, in effect, a d.c. component made up of the difference between the positive and negative amplitudes of the sine wave. This will introduce noise in the recording and impart some 'wasted' remanence to the tape.

It is usual for the oscillator valve to provide both bias and erase currents. In most of the commercial tape decks, the combined record-playback head is usually of high impedance, while the impedance of the erase head may be high or low. To accommodate low-impedance erase heads, a separate oscillator output, in the form of a secondary winding to the oscillator coil, is required for matching purposes.

Because the characteristics of tape-recorder heads are not standardised, it is impossible to design an oscillator coil that will suit all types of head. Some degree of flexibility is possible if a tapped secondary winding is used, but the oscillator stage of a tape amplifier circuit may have to be adapted to a greater or lesser extent in accordance with the recommendations of manufacturers of tape heads and oscillator coils.

FREQUENCY RESPONSE

The degree of magnetisation of the tape which results from recording with a current through the head which is constant at all frequencies is of the form shown by the curve ABC of Fig. 11. The reduced intensity in the treble region is attributable in the main to the self-demagnetisation of the tape.

To outline the recording process briefly – the tape is drawn steadily past the recording head, and the field produced by the signal current in the head induces some degree of residual magnetism in the section of the tape nearest to the head. The amplitude of the signal governs the amount of magnetism, and the sense (whether positive or negative) controls the direction of the magnetism in each section.

On the basis of the molecular theory of magnetism, the 'molecular' magnets in the tape are aligned by the magnetising field: the strength of the field governs how many of these magnets are brought into line; the direction of the field dictates whether the alignment is with north or south poles leading.

The field set up by a sinusoidal signal current in the head would change direction regularly, so that the pattern of magnetism included in the tape would have adjacent sections of the tape with north and south poles leading alternately (see Fig. 12). Thus, the poles of each section of the tape would be adjacent to the like poles of the neighbouring sections.

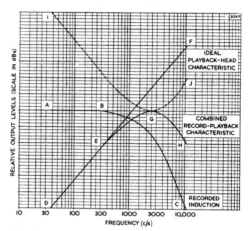

Fig. 11—Tape record-playback characteristics

The length of these sectional magnets in the tape depends on the speed at which the tape passes the head, and on the frequency at which the magnetising field changes direction. For a constant transport speed, the length varies inversely as the frequency of the signal. For low-frequency signals, the magnets are comparatively long, but for high frequencies they are short.

The ability of the tape to resist demagnetisation depends, in part, on the length of these sectional magnets. If they are short, the opposing fields set up by the neighbouring like poles will cause misalignment more readily than if they are long. Hence the demagnetising effect is greater at the higher frequencies than at the lower ones, and consequently there will be less residual magnetism at the higher frequencies even if the signal amplitudes are constant throughout the frequency range.

For the playback process, the magnetised tape is drawn past the playback head at a constant speed, and the magnetic fields associated with the sections of the tape move and cut the coils of the head, thereby setting up voltages in the coils. The rate of change of the flux cutting the coils governs the voltage generated. Thus, the rapidly changing fields set up by the sections of the tape on which high-frequency sounds have been recorded must of necessity give rise to larger voltages than the sections containing low-frequency impressions. This is so, even if the impressions at both high and low frequencies have the same intensity of

DISTORTION FROM TAPE EQUIPMENT

magnetisation. The actual speed of the tape has no bearing on the relative levels of the output at the various frequencies. An increase in the transport speed simply multiplies all levels by the same amount.

The rate at which the playback voltage rises with frequency is equivalent to 6dB per octave. In Fig. 11, the line DEF indicates the frequency response of the playback head only. (A tape which has been ideally magnetised is assumed for this response curve.) The response curve for the recording head alone will depend on various practical considerations. For example, a tape made of a magnetic material of high coercivity is difficult to demagnetise, and if one is used, then the demagnetisation losses at high frequencies will be less than if a low-coercivity tape is used. The use of a high coercivity tape, of course, makes intentional erasing more difficult.

The response for the complete equipment, (that is, the response combining both recording and playback deviations) is indicated by the curve DEGH of Fig. 11. To obtain an equalised output from this response curve it is necessary to introduce boost at both ends of the frequency range. For ideal equalisation, the compensating curve must be a mirror image of the curve DEGH, so that the equalising response curve needs to be the curve IGJ.

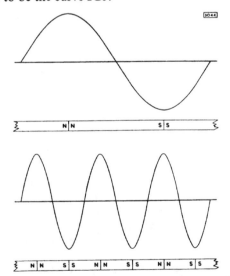

Fig. 12—*Lengths of sectional magnets corresponding to low- and high-frequency signals*

There will be, in fact, further high-frequency losses encountered in tape recording. The tendency of the a.c. bias signal to act as an erasing signal during the recording process is more pronounced at high frequencies because the remanence is less deeply seated at these frequencies. Also, the physical dimensions of the gap in the playback head cause losses when the size of the sectional magnets and the gap are commensurate. (This is usually referred to as 'gap effect'.) These combine to produce a more accentuated drop in the response curve in the treble region.

Unfortunately, equalisation is not simply a matter of compensating for bass and treble deviations. Several performance requirements are in conflict, and a compromise is the best that can be achieved. A wide frequency response, low distortion or a high signal-to-noise ratio can each be obtained only at the expense of the others. Over a restricted range, an increase in the bias current, for example, lessens distortion, but causes treble attenuation and tends to lessen the signal-to-noise ratio. Furthermore, equalisation is a function of the physical properties of the tape used. The equalisation provided with any combination of tape deck and amplifier may, for example, provide a level response over a wide frequency range with one brand of tape, but may give rise to a large treble peak with another brand.

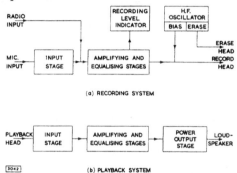

Fig. 13—*Block schematic diagram of tape recording and playback systems*

EQUALISATION

A tape amplifier can be represented essentially by the block schematic diagram of Fig. 13. Normally, the design of both amplifying units will follow general a.f. practice, although attention must be paid to certain requirements of the equipment. Precautions will be required to minimise hum pick-up, and the overall signal-to-noise ratio should be at least 40dB.

The attenuation of the bass response in a tape recorder takes place mainly in the playback head. The treble response suffers its greatest losses in the recording process. Thus, in order to load the tape more or less evenly, it is best to provide treble boost in the recording amplifier and to apply bass boost in the playback amplifier. Some treble accentuation may also be required in the playback amplifier, but this has to be limited in order to avoid much amplification of background noise.

CHAPTER 3
High-quality Amplification

The principal features of good amplifying equipment can be summarised as follows:

1. The distortion produced by the amplifier should be negligible up to the maximum output level. Distortion is the presence in the output of frequency components which were not present in the input signal. These consist, in the first place, of harmonics of the original frequencies and, in the second, of sum and difference tones caused by intermodulation between different frequencies.

2. The response of the amplifier should be uniform throughout the audible frequency range. The average ear will respond to frequencies in the range of 30c/s to 15kc/s. The upper limit of this range may extend to 20kc/s, particularly with young people, or be well below 15kc/s with older people. The upper partials of some musical instruments extend to about 10kc/s and the sound spectrum associated with them to about the upper limit of the audible range. To make realistic reproduction possible, therefore, the amplifier should handle a range of frequencies at least as wide as that which can be heard.

3. The response of the amplifier to signals of a transient nature should be good. Many sounds, particularly those deriving from musical instruments, rise very rapidly to a high intensity and decay relatively slowly. Such sounds are spoken of as 'transients', and examples of these are the sounds resulting from the clashing of cymbals or the plucking of strings.

   The steeply rising wave fronts of transient sounds can be shown to consist of a wide range of component frequencies. The ability of an amplifier to reproduce these transients faithfully will therefore depend on two things: First, the frequency response of the system must be wide and, second, the phase shift over the whole frequency range must be small. Variations in the relative phasing of the component frequencies of a transient would result in a change in its aural character.

4. An adequate reserve of output power should be available. For faithful reproduction, the sound level should be comparable with that of 'live' conditions. The amplifier should thus be capable of handling peak powers considerably above the average level to allow peak sounds to be reproduced without overloading and audible distortion.

5. The output resistance of the amplifier should be low. This will improve the performance of the loudspeaker and ensure crisp and clean reproduction, particularly of transients. Air-loading of the loudspeaker limits, to a large extent, the low-frequency resonance of the cone and suspension. The electromagnetic damping afforded by a low output resistance in the amplifier is, however, effective in ensuring adequate control of the cone movement over the whole frequency range.

   The output resistance should preferably be much less than the impedance of the loudspeaker voice coil, the ratio of the two being termed the 'damping factor'. In practice, a damping factor of above 10 is desirable.

6. The level of hum and noise should be low: the presence of these in the loudspeaker output will detract from the enjoyment of otherwise good reproduction. Thus the amplifier should not give rise to noise or hum at an audible level.

PERFORMANCE OF AMPLIFYING EQUIPMENT

The performance of amplifying equipment is normally stipulated with reference to some or all of the points enumerated above, and the interpretations to be given to the values quoted in subsequent chapters of this book are discussed in the following paragraphs. In the chapters on stereophonic equipment, the performance of each channel is quoted. Combined values for the two channels can be misleading. For example, a reserve of output power (see below) of, say, 5W per channel does not mean that an output of 10W will be delivered before overloading occurs. Overloading will occur in the left-hand channel if that channel has to deliver more than 5W, and similarly in the right-hand channel. For sound sources placed symmetrically between the right- and left-hand microphones, the total power reserve will be 10W, but for sounds originating in either extreme left- or right-hand positions, the total reserve is little more than the reserve of the left- or right-hand channel.

Power Reserve

The audio power available at the output of an amplifier is defined as $(V_{load})^2/R_{load}$, where V_{load} is the voltage developed across a load resistance R_{load} connected to the output terminals of the amplifier. The rated output power, or the power reserve, of the amplifier is the maximum audio power which can be obtained without exceeding either the limiting values of the valves or the level of distortion permitted in the specification for the equipment.

Distortion

The principal form of distortion occurring in the output from amplifying equipment is non-linear distortion which is normally divided into harmonic, intermodulation and beat-note distortion. Each subdivision contributes some power to the amplifier output at frequencies which differ from those occurring in the input signal.

Harmonic Distortion

Power which occurs in the output at second, third, fourth, and so on, harmonics of the fundamental signal frequency comprises harmonic distortion. It is expressed as a percentage ratio of the power associated with the particular harmonic to the total output power of the amplifier. Total harmonic distortion is the ratio of the power associated with all the harmonics to the total output power. The total harmonic distortion D_{tot} is the r.m.s. value of the individual distortions D_2, D_3, D_4, etc. – that is:

$$D_{tot} = \sqrt{(D_2^2 + D_3^2 + D_4 + \ldots)}$$

Intermodulation Distortion

If an input signal contains several different frequencies, any non-linearity in the amplifier will give rise to modulated waveforms having frequencies which are the sums and differences of the interacting waveforms. The extent of this distortion is assessed by measuring the degree of interaction between two pairs of test signals. The interaction between signals of very different frequencies is quoted as intermodulation distortion and that between signals of nearly equal frequencies is given as beat-note distortion.

Intermodulation distortion is measured between test signals having frequencies of 40c/s and 10kc/s. The ratio of the peak-to-peak amplitudes of the l.f. and h.f. signals is 4 : 1. The output obtained with the two signals is taken to be equivalent to the output obtained with a single sine-wave signal the peak-to-peak amplitude of which is equal to the peak-to-peak amplitude of the combined waveform. The distortion is quoted as the r.m.s. value of the amplitudes of the sum and the difference waveforms, expressed as a percentage of the amplitude of the h.f. signal.

Beat-note distortion is measured with two test signals having frequencies of 14 and 15kc/s. The amplitudes of these signals are equal, and the resulting difference waveform, expressed as a percentage of the test signal amplitudes, gives the beat-note distortion. The equivalent output power of the combined waveform is defined as for intermodulation distortion.

The test frequencies given above are British Standard test frequencies. American Standard test frequencies are 70c/s and 7kc/s for intermodulation distortion and 9 and 10kc/s for beat-note distortion. Distortion figures would be lower when referred to the American Standard test signals than when referred to the British Standard signals, but the actual distortion is obviously the same in the same amplifier. Examples of this are shown in Fig. 12 on page 51 for the 10W amplifier.

Hum and Noise

Contributions to the output from various stray signals picked up at points in the amplifier are normally assessed together, and are often measured as the voltage developed at the output when the input is short-circuited to earth. This voltage is expressed in decibels as a fraction of the rated output voltage measures across the load resistance, so that:

Hum and Noise (in dB)
$$= 20 \log_{10}\left\{\frac{\text{voltage with input closed}}{\text{rated output voltage}}\right\}$$

A level of hum and noise of, say, 60dB, means that the rated output voltage is 1000 times the voltage developed when the input is short-circuited to earth.

Negative Feedback

Negative voltage feedback is used in high-quality amplifiers to improve the performance. Part of the signal is taken back and injected in an earlier stage of the amplifier in opposite phase (180° out of phase)

OUTPUT STAGES

thus reducing sensitivity. It is usual to refer to the amount of feedback in terms of the ratio of voltage gain of the amplifier without feedback to the gain with feedback. Thus feedback of, say, 26dB would mean that the gain without feedback is 20 times the gain with feedback.

The gain of an amplifier without feedback must therefore be great enough to allow for the reduction in gain caused when the feedback is applied. This drawback is outweighed by the following advantages to be obtained from using feedback: (1) reduced distortion; (2) wider and flatter frequency response; (3) reduced output impedance; (4) reduced phase shift; (5) less dependence on small changes in supply voltages, etc.

Circuit Design

Although the power handling capacity of an audio amplifier is not the property which is most important to the listener – a low level of distortion is usually considered to be so – it is nevertheless the prime concern of the circuit designer. The reserve of output power required for realistic reproduction of orchestral music depends mainly on the size and acoustical nature of the room and, to a lesser extent, on the taste of the listener. In the home it is generally considered that a peak output power of 7 to 10 watts will be adequate (assuming a loudspeaker efficiency of 5%). If simplicity and economy are the governing considerations, a reserve of 3 watts can, if the amplifier is carefully designed, give a generally acceptable standard of performance. In large rooms or small assembly halls, however, the conditions will probably merit a maximum output power of at least 20 watts. The type of output stage used will depend on the maximum power demanded of the amplifier. Consequently, the design of amplifying equipment will normally proceed from output to input, the requirements of the output stage dictating to a large extent the design of preceding stages.

OUTPUT STAGES

For power reserves of up to about 5 watts, single-valve output stages in which the valve operates under Class A* conditions, are considered satisfactory. Triode valves used as power amplifiers yield a relatively distortion-free output, but their power-handling capacity is generally inadequate. The power-handling capacity of pentodes, however, is sufficient to permit their use in such stages and although the inherent distortion is high, negative feedback can be used to reduce it to an acceptable level.

There exists a choice of two basic forms of output stage from which a good-quality output of more than 5 watts can be delivered to the voice-coil of a loudspeaker. The two well-known forms are:

1. A Class AB† push-pull pentode (or tetrode) output stage;
2. A Class A or AB push-pull triode output stage.

The choice between these is largely a balance between economy and performance.

Pentode Push-pull Stages

The use of pentodes of the 12W anode-dissipation type operating in a conventional Class AB push-pull stage enables an effective output of 12 to 13W to be obtained easily assuming an output transformer efficiency of about 80% (which is a fairly conservative estimate for the efficiency). The appropriate supply voltage, limited by valve ratings, is about 300 to 320V. The power efficiency (that is, the ratio between the audio output power and the d.c. input power) of such a stage is fairly high, being 50%, for example, in a typical stage using two Mullard output pentodes, type EL84. Harmonic distortion, however, is of the order of 3 to 4% at full output and, thus, a large amount of negative feedback is necessary to reduce the distortion to an acceptably low level (say 0·5%) at maximum output.

The conditions for Class AB operation normally recommended and published by valve manufacturers are based on measurements made with continuous sine-wave drive. The bias under zero-drive conditions and the anode-to-anode load resistance are so chosen that the optimum performance is achieved when the working points of the valves are displaced under driven conditions. This displacement is caused by the effect of increased anode and screen-grid currents on the cathode-bias circuit. For a typical output stage working from a 310V mains supply and using EL84s, the rise in cathode current – and thus in cathode bias voltage – with a sinusoidal signal voltage is about 40% at full drive.

When such a stage is used in the reproduction of speech or music, however, the operating conditions are different. The mean amplitude of the input signal is now very small compared with the peak values which occur from time to time, and thus the mean

---

*Class A operation is that in which the values of bias and signal voltages applied to the control grid of the valve ensure that anode current always flows.

†Class AB operation is that in which the values of bias and signal voltages applied to the grids of the valves cause anode current to flow in each valve for appreciably more than half, but less than a whole, cycle of the signal voltage.

variation in cathode current is also very small. Because of the relatively long time constant of the bias network, even under peak signal conditions, the displacement of the working point is small enough for the stage to be considered as working with fixed bias. When a nominal, cathode-biased, Class AB stage is examined under the corresponding fixed-bias conditions with a sine-wave input, it is found that, at high output levels, distortion is greater than when cathode bias is used. These two conditions are illustrated for EL84s by Curves 1 and 2 in Fig. 1. The quiescent bias is the same in both cases, Curve 2 showing normal operation with cathode biasing and Curve 1 showing operation with fixed bias. The results indicate that, in practice, a cathode-biased Class AB stage designed on the basis of a sinusoidal drive will produce increased distortion when peak passages of speech or music are being reproduced.

One practical method of improving performance is to adjust the quiescent operating conditions in the output stage so that they are nearly optimum for fixed-bias working, although cathode bias is still being used. This entails a smaller standing current $I_{a(o)}$ and a lower anode-to-anode load impedance R_{a-a}. These changes result in larger variations in the instantaneous anode and screen-grid currents when the stage is driven, but the effect of these is lessened because the time constant of the cathode-bias network has also been increased. The excursion of the working bias is still kept very small under driven conditions.

It is found that good short-term regulation of the h.t. voltage is ensured by the use of large (50µF) electrolytic capacitors for the anode and screen-grid supplies. Peak currents corresponding to near-overload conditions are effectively provided by the capacitors with a reduction in the line voltage of well under 0.5%, and the instantaneous power-handling capacity of the stage is not impaired.

This modified design is described more fully in Chapter 6 as an alternative 'low-loading' version of the 10W amplifier, and it has proved to be a very satisfactory arrangement. A secondary feature of the use of these operating conditions is that the 12W output valves each run at a mean dissipation of only 7.5W. The corresponding fixed bias conditions in this case are represented by Curve 3 in Fig. 1.

The low-loading form of operation is, however, suitable only for use in the reproduction of speech or music and cannot be used with sine-wave input without excessive distortion resulting. For this reason it is difficult to measure directly the distortion levels which hold under practical conditions.

A second method of improving the performance of pentode push-pull output stages is the use of 'distributed-load' conditions. This is discussed in detail later in the chapter. Depending on the precise loading used, the variation in anode and screen-grid currents can be reduced to such a level that almost identical performance is obtained with cathode and fixed bias.

Triode Push-pull Stages

The level of distortion is inherently low in a triode push-pull output stage in which the valves operate more or less under Class A conditions. It is found that if 25W pentodes (or tetrodes) are connected as triodes, an output power of 12 to 15W with a level of harmonic distortion below 1% can be obtained using a supply voltage of 430 to 450V.

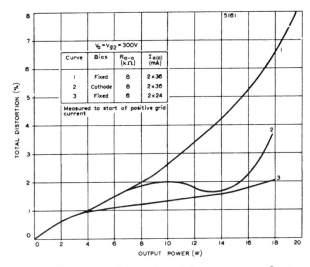

Fig. 1—Variations of distortion with output power for two EL84s in class AB push-pull operation

Maximum output power and the corresponding distortion vary appreciably with the value of load impedance. The curves of Fig. 2 illustrate the performance of two Mullard high-slope output pentodes, type EL34, connected as triodes in a push-pull output stage, operating slightly below their anode-dissipation rating of 25W. For anode-to-anode load impedances below 7kΩ, either a shared, bypassed cathode resistor or separate bypassed cathode resistors can be used. Above 7kΩ however, improved operation is obtained if a shared resistor with no bypass capacitor is used. Operating conditions approach Class A as the anode-to-anode impedance is raised, and the optimum performance for high-quality output stages is obtained with a load impedance of about 10kΩ. An output of 14W is then delivered by the valves with total harmonic distortion well below 1%.

OUTPUT STAGES

This type of output stage found favour for a number of years in high-quality amplifiers giving about 12W effective output. Because of the low inherent distortion of these triode-connected stages, less negative feedback is required to give acceptable linearity than has to be used in pentode or tetrode stages giving similar output powers. Furthermore, in a three-or four-stage design in which feedback is applied over the whole amplifier, including the output transformer, it is thus possible to obtain greater margins of stability for a given level of distortion.

Distributed Loading

The conditions of distributed loading are achieved by applying negative feedback in the output stage itself. In the simplest form of distributed-load operation, the screen grids are fed from suitably positioned taps on the primary winding of the output transformer, and the stage can be considered as one in which negative feedback is applied in a non-linear manner via the screen grids.

The characteristics of the output stage under these conditions of operation lie between those for pentode and triode operation. They approach the triode characteristics as the percentage of the primary winding common to the anode and screen-grid circuits increases. Under optimum conditions, about two-thirds of the power-handling capacity of the corresponding pentode stage can be realised with a much lower level of distortion while, at power levels corresponding to triode operation, the distortion also is of the order corresponding to triode operation. At the same time, the output impedance is reduced to a level approaching that obtained if a conventional triode push-pull stage is used.

An output stage with distributed loading can thus be used with pentodes of the 25W type in high-quality amplifiers designed for output powers well in excess of 15W, the power efficiency being appreciably greater than with triode operation. Alternatively, the performance of 12W pentodes can be improved considerably and, although the power handling capacity is reduced somewhat, effective output powers of 10 to 12W can still be obtained.

A comparison is given in Table 1 of triode, pentode and distributed-load operation for the Mullard output pentodes, types EL34 and EL84. For the EL34, the comparison between distributed-load operation and triode operation is of most interest. It will be seen from this that distributed-load operation using a tapped-primary output transformer enables the power-handling capacity to be more than double that

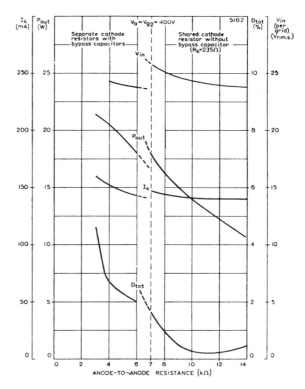

Fig. 2—Performance characteristics of two EL34s in triode-connected push-pull arrangement

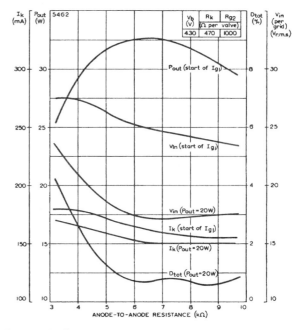

Fig. 3—Performance characteristics of two EL34s in pentode-connected push-pull arrangement under conditions of distributed load

OUTPUT STAGES

possible with triode operation whilst, at the same time, distortion in the stage can be kept very low.

Although with a common-winding ratio of 0·2 (that is, with 20% of the primary winding common to the anode and screen-grid circuits) the distortion level is comparable with triode conditions, it has been found that appreciable improvement is obtained at higher output powers if the ratio is further increased. Progressive improvement has been obtained as the percentage of common primary winding is increased up to 40 or 45%. The power-handling capacity of the stage is reduced further as the ratio is increased, but at least 35W can be obtained with a level of distortion at the onset of grid current of about 2·5%.

Typical performance curves of the EL34 when used with an output transformer having the primary winding tapped at 43% of the turns are shown in Fig. 3. The output powers quoted are those delivered to the load in the secondary circuit.

With valves of the 12W dissipation class, comparison with normal pentode operation is more significant. Appreciable reduction in odd-harmonic distortion is again obtained under distributed-load conditions, and an output of approximately 15W is delivered by the valves if the common winding ratio is 0·2.

There are two advantages not apparent in Table 1 in using a common-winding ratio of about 0·4 when a high output power is available. First, almost identical performance figures are obtained under cathode- and fixed-bias conditions since, with the closer approach to Class A triode operation, variations in anode and screen-grid currents are reduced when the stage is driven. Second, as with normal triode operation, output power and distortion are less dependent on the value of load impedances: with a common-winding ratio of 0·4, little change in performance is produced by altering the anode-to-anode load from 6 to 9kΩ. However, Table 1 shows that there is little benefit to be achieved in respect of distortion by increasing the common-winding ratio

TABLE 1

Comparison between Triode, Pentode and Distributed-load Operaton of EL34s and EL84s

Valve	Mode of Operation	Operating Conditions					Total Distortion (%)				
		V_a (V)	V_{g2} (V)	R_k (Ω)	R_{a-a} (kΩ)	R_{g2} (Ω)	10W	14W	20W	30W	40W
EL34	Triode connection	400	*	470 (per valve)	10	*	0·5	0·7	—	—	—
	Distributed load										
	(a) 20% common winding	400	400	470 (per valve)	7·0	1000 (per valve)	0·7	0·8	1·0	1·5	5·0
	(b) 43% common winding	400	400	470 (per valve)	6·6	1000 (per valve)	0·6	0·7	0·8	1·0	—
	Pentode connection	330	330	130 (common)	3·4	470 (common)	1·5	2·0	2·5	4·0	6·0
							5W		10W		15W
EL84	Triode connection	300	*	150 (common)	10	*	1·0		—		—
	Distributed load										
	(a) 20% common winding	300	300	270 (per valve)	6·6	—	0·8		1·0		1·5
	(b) 43% common winding	300	300	270 (per valve)	8·0	—	0·7		0·9		—
	Pentode connection	300	300	270 (per valve)	8·0	—	1·5		2·0		2·0

*Screen grid strapped to anode

INTERMEDIATE STAGES

beyond 0·2 and, because of the greater power-handling capacities possible, the circuits described in this book which use distributed loading are designed for output transformers having 20% of the primary winding common to the anode and screen-grid circuits.

INTERMEDIATE STAGES

To ensure stability in an amplifier when negative feedback is used to reduce distortion, the phase shift must be small. Thus, the number of stages should be a minimum. For low-output equipment such as the 3W amplifier described in Chapter 7, it is usually possible to obtain the desired output with only a voltage-amplifying input stage and a power-amplifying output stage. However, for equipment using a push-pull output stage, the penultimate stage must be capable of providing a well-balanced push-pull drive voltage of adequate amplitude. Consequently, extra stages will be required to give the push-pull drive voltage and to provide sufficient amplification of this voltage.

With push-pull stages using 25W pentodes such as the EL34, the maximum drive voltage required is approximately $2 \times 25 V_{r.m.s.}$. For stages using 12W pentodes (the EL84, for example) the requirement is about $2 \times 10 V_{r.m.s.}$. (These input-voltage requirements are similar for triode, pentode or distributed-load operation.) If a high-impedance double triode such as the Mullard ECC83 is connected as a cathode-coupled phase splitter, it is possible to produce the above drive voltages and also to perform the necessary phase splitting in a single intermediate stage.

A cathode-coupled phase splitter using an ECC83 is shown in Fig. 4. A negative pulse at the grid of the first triode section of the ECC83 produces a positive pulse at the first anode, so that V_{out} is positive-going. The cathodes are strapped together, and therefore go negative at the same time. The negative pulse on the second cathode is equivalent to a positive pulse on the second grid so that V'_{out} is negative-going. Hence V'_{out} is 180° out of phase with V_{out}. The grid of the second section of the valve is capacitively earthed by C3 for cathode input, and R2 ensures correct d.c. conditions in the second section.

The input grid of the circuit is directly coupled to the anode of the preceding valve so that the grid is at a direct potential of between 70 and 80V. The bias for the valve is obtained by the voltage-drop across the shared cathode resistor R5, and the direct voltage of the cathode is therefore high. The cathode currents of both sections flow through R5, the d.c. components being additive and the a.c. components in opposition.

For there to be some a.c. signal remaining on the cathode to inject into the second section, the two sections cannot be operated with exactly equal anode loads nor with equal grid resistors in the following stages. The anode load R4 of the earthed triode section should be slightly higher than R3 to give perfect balance. Because of the high amplification factor of the ECC83 (μ is 100) exactly equal resistors cannot cause more than 3% lack of balance and nominally equal resistors, matched to within 5%, result in a lack of balance which cannot exceed 2%.

The grid resistors R6 and R7 for the output valves must be of close tolerance ($\pm 5\%$) as they form part of the load to the phase splitter. The frequency at which low-frequency unbalance occurs depends on the time constant $R_2 C_3$ and this can be made long enough to maintain adequate balance down to low audio frequencies.* High-frequency balance is largely determined by the layout of the wiring which can make shunt capacitances unequal. The inter-electrode capacitances of the ECC83 are sufficiently small and equal between the two sections for their shunting effect on the circuit to give negligible unbalance at high frequencies.

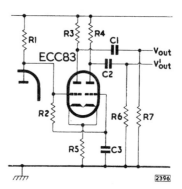

Fig. 4—Cathode-coupled phase splitter

The effective voltage gain (V_{out}/V_{in} or V'_{out}/V_{in}) is about half that of one section used as a normal voltage amplifier. Nevertheless, there is sufficient gain in this stage because of the high amplification factor of the ECC83. The absence of the input coupling capacitor reduces low-frequency phase shift, and makes for good stability when feedback is applied.

Because of the direct coupling to the preceding stage, the anode voltage of the preceding valve determines the operating conditions of the ECC83. If this anode voltage is too high, the negative bias on the phase splitter is too low, and overdrive on peak signals will result in grid-current distortion. If the anode

---

*R2 will be of the order of 1MΩ so that unbalance is caused if any leakage current flows through C3. This is a common cause of distortion.

voltage is too low, the bias of the phase splitter will be too high and the valve will operate away from the linear part of its dynamic characteristic, and distortion will be greater.

An alternative type of phase splitter – or more correctly, phase reverser – is illustrated in Fig. 5. This circuit has various names such as anode follower, see-saw and paraphase.

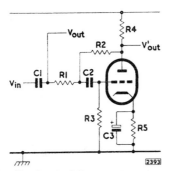

Fig. 5—Anode-follower phase splitter

One input for the push-pull output stage is the output V_{out} from the stage preceding the phase reverser, and the other is the output V'_{out} from the phase reverser. The input for the phase reverser is taken from the potential divider R1,R2 connected between the two outputs. When V_{out} is positive-going,

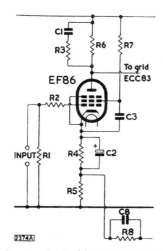

Fig. 6—Typical EF86 input stage

the grid is positive-going, and the anode is negative-going, so that V_{out} and V'_{out} are 180° out of phase. The values of R1 and R2 are chosen so that the gain of the phase-reversing stage is unity (that is, $V'_{out}/V_{in} = 1$). The ratio R_1/R_2 is equal to the ratio $(A-1)/(A+1)$ where A is the gain of the phase-reversing valve measured from grid to anode.

The circuit is less widely used than the phase splitter described above because a lack of balance between the two outputs can result from the coupling capacitors. However the circuit has two advantages:

1. It does not require a voltage difference between heater and cathode, so that it is suitable for directly-heated valves or for double-valves in which the second sections are needed for other purposes.

2. If a voltage does exist between heater and cathode – as in the heater chain of a d.c./a.c. amplifier for instance – the large amount of negative feedback from anode to grid can be used to counteract excessive hum pick-up.

INPUT STAGES

A low level of hum and noise is essential in the input stage of an amplifier because high gain is required in this stage to provide a good signal-to-noise ratio and to compensate for the reduction in gain resulting from the use of negative feedback. The recommended valve for input stages is the Mullard low-noise pentode, type EF86 (or, in d.c. equipment with series heater chains, type UF86). A conventional EF86 voltage-amplifying stage is drawn in Fig. 6.

The anode load is shown shunted by a CR network which produces an advance in phase and thus increases the stability of the amplifier at high frequencies. Negative feedback (from the secondary winding of the output transformer) is introduced across a high stability resistor R5 in the cathode circuit. The feedback resistor R8 should be of the high-stability type. The feedback network C8,R8 attenuates frequencies in the ultrasonic range and again gives increased stability. The stage is intended for direct coupling to the input grid of a cathode-coupled phase splitter so that the phase shift at low frequencies is minimised and the low-frequency stability of the amplifier with feedback is improved.

The following sections describe the precautions needed to ensure that the good properties of the EF86 are realised in the typical circuit of Fig. 6.

Noise

High stability, cracked-carbon resistors should be used for the anode load R6, the screen-grid resistor R7, and the cathode-circuit resistors R4 and R5. If normal carbon resistors are used, the low noise level of the EF86 may be masked by additional non-thermal noise generated by the flow of current in the resistors.

The low-frequency noise generated by the EF86 itself is most conveniently specified as an equivalent

INPUT STAGES

voltage at the control grid. For an h.t. line voltage of 250V and an anode load of 100kΩ, the equivalent noise voltage is 2μV for a bandwidth of 25c/s to 10kc/s. The bandwidth has to be specified because the random noise of the valve covers all possible frequencies, and also because the noise voltage is more or less proportional to the bandwidth. For comparison, a resistance of 100kΩ in the grid circuit produces a thermal noise voltage of about 4μV for the same frequency range.

Hum

The cathode resistor R4 is bypassed at signal and hum frequencies by the capacitor C2. Heater emission or the choice of an unsuitable valveholder leading to capacitive coupling or leakage between the pins, may give rise to a heater-to-cathode current. The following list indicates points which should be considered when constructing an amplifier:

1. A low-loss valveholder must be used. A nylon-loaded phenolic holder, preferably skirted, is generally adequate.

2. The external leads to the heater should be run as a twisted pair to neutralise their external magnetic fields. (The heater of the EF86 is itself made in the form of a spiral.) The heater wiring should be kept as close to the chassis as possible.

3. Earth returns should be made to a busbar. Connections can be made to the central spigot of each valveholder in the first instance. The busbar should only be connected to the chassis at one point, preferably near to the input terminal. Hum picked up from currents circulating in the chassis is then kept to a minimum. Wherever possible, an external earth connection should be made.

4. Where a recommended layout for the components is given, there will be no danger of hum being generated by the magnetic field of the mains transformer acting directly on the electrode supports. If some other layout is tried, the possibility of this form of hum occurring should be considered very carefully.

When used as a normal voltage amplifier with an h.t. line voltage of 250V, an anode load of 100kΩ and a grid resistor of 470kΩ, the maximum hum voltage of the EF86 itself when one side of the heater is earthed, is about 5μV at the grid, and the average value is about 3μV. The maximum level with a centre-tapped heater is only 1·5μV. Additional external screening of the valve is not normally required.

Microphony

The EF86 has a rigid electrode structure. When correctly sited, the action of sound waves (acoustic feedback) and vibration on the electrode structure will not produce any audible howl. When the valve is sited in an obviously doubtful or unsuitable position as, for example, in tape recorders, a flexible mounting for the valveholder or a separate weighted sub-chassis is advisable. There are no appreciable resonances in the EF86 below 1kc/s. At higher frequencies, the effect of vibration is generally negligible because of the damping provided by the chassis and valveholder.

Starvation Operation

In a single-valve output stage, pentodes are essential if adequate output power is to be obtained. Inherent distortion with pentodes, however, is high, and a large amount of negative feedback is needed to reduce the distortion to an acceptable level. Consequently, a very high gain is essential in the input stage to compensate for this, particularly if only one stage of voltage amplification is used.

It is well known that the gain of a voltage amplifier can never reach the theoretical maximum represented by the amplification factor μ of the valve. The gain, in fact, is given by the expression $\mu R_a/(R_a+r_a)$, where r_a and R_a are the internal slope resistance of the valve and external load resistance respectively.

Practical considerations in conventional circuits restrict the extent to which the anode load can be increased. For example, the maximum will depend on the effect of shunt capacitances and the need to maintain the frequency response at the upper end of the audio spectrum. The load will normally be less than 500kΩ and this value, used with an EF86 operating with a line voltage between 250 and 350V, will give a stage gain of about 250. In these conditions, the anode and screen-grid voltages will be 60 and 70V respectively.

If a more severe limitation of frequency response is acceptable, much higher values of anode load may be used. These higher values produce 'starvation' operating conditions: valve currents and voltages are much smaller than those of conventional stages, but the stage gain is much greater. The frequency response is restricted under starvation conditions, but it can be extended considerably by the use of corrective feedback.

Because of the high value of anode load for starvation conditions, the input impedance of the following stage will need to be high. Preferably, it should not be less than 10MΩ. Direct coupling to the grid of the following stage is possible, and the screen grid of the voltage amplifier can be fed from the cathode circuit of the following stage. Because of the low values of current and voltage and the favourable effect of these on negative grid current, starvation conditions are useful in high input-impedance voltage amplifiers.

PRE-AMPLIFIER STAGES

Pre-amplification may be required with power amplifiers to provide extra voltage gain when certain kinds of input source are used. Often, it is convenient to include the treble and bass tone controls and the volume control in the pre-amplifier. Also, the pre-amplifier can be designed to give compensation for the bass attenuation and treble boost which have been applied by the manufacturers during recording to obtain more favourable ratios of signal-to-motor-rumble and signal-to-surface-noise. Pre-amplifiers for use with tape-recording equipment will also provide equalisation for the inherently non-linear nature of magnetic recording.

Pre-amplifiers for stereophonic equipment must necessarily be more complex than monaural units because a stereophonic pre-amplifier is more or less two monaural circuits in one. Furthermore, coupling must exist between corresponding controls in each section of the stereophonic circuit so that comparable adjustments can be made to both channels. If the principle of design adopted is that both channels must be identical, then rigid ganging of the controls can be used. However, if it is intended to cater for non-identical channels,* concentrically-operated controls which will permit individual and coupled adjustment of the channels, will be needed.

---

*There is strong evidence suggesting that the two channels of a stereophonic arrangement need not be identical. Excellent results are obtainable, for example, if one channel uses a 20W power amplifier and the other, a control-less 10W amplifier. Of course, the settings of the gain and tone controls in each channel of the pre-amplifier will need to be different because of the different characteristics of the power amplifiers.

Even if nominally identical channels are used, the acoustical outputs from the two loudspeakers will not be exactly the same unless precautions are taken. Differences can occur because of (i) the difference in the outputs from the halves of stereophonic pick-up heads, (ii) the unequal sensitivities of the loudspeakers and (iii) the very small differences in gain of the two amplifying channels. If the volume control consists of a dual-concentric potentiometer, individual adjustment to each channel will rectify any lack of balance. If a dual-ganged potentiometer is used, a special balance control is required. It should be possible with this control to increase the gain in one channel while simultaneously decreasing it in the other. Also, it should be possible to increase and decrease the gain of one channel with respect to the other, and it is desirable that the degree of control available in either sense should be the same, so that a symmetrical, 'centre-zero' arrangement is needed.

Another facility normally required in a stereophonic pre-amplifier which is not necessary in a monaural circuit is a switch to allow the transference of the input signals from one channel to the other. Also, this switch usually serves to combine both channels, so that the equipment can be used with monaural recordings.

All pre-amplifiers, because they provide the inputs to the power amplifiers, must be designed to send as little hum and noise as possible into the main amplifier. Hum and noise in the early stages count as signal voltages and are not reduced by feedback. There is therefore no advantage in including a pre-amplifier in the feedback loop. In fact, the switching required for compensating and equalising circuits and the frequency-selective networks in a high-gain pre-amplifier are likely to produce large phase shifts and, consequently, to increase the risk of instability if included in the feedback loop.

CHAPTER 4

General Notes on Construction and Assembly

Details are given in subsequent chapters of the dimensions, layout and wiring of each item of equipment. This information is extensive, but no claim is made that it is exhaustive. The constructional sections of each chapter should be regarded principally as giving some guidance for assembling the equipment, but should not be considered to be absolutely inviolate sets of rules and instructions. An intelligent use of these sections, coupled with references to circuit diagrams, should produce excellent results.

The suggested arrangements of components in the various chassis are those adopted in the prototype equipment. The convenience of the constructor has been borne in mind in evolving these layouts, and economy of space has been effected whenever possible. However, the underlying principle has always been to obtain the best possible standard of performance for a given expenditure. A good margin of stability, a low level of hum, and so on, have never been sacrificed in an attempt to achieve attractive, but less fundamental, properties such as miniaturisation.

There is no theoretical method of deducing the best layout for audio equipment. It has to be found experimentally, and it follows that the effect of any deviations from an established arrangement cannot be predicted. Consequently, any alterations to layouts recommended in this book should not be undertaken lightly. They should be dictated by strong reasons, and the subsequent variations in performance should be analysed closely.

Similarly, recommendations in this book for good-quality valveholders and for high-stability resistors, for example, are based on experimental proof of their necessity. Any change from the recommendations can alter the performance considerably. The voltage ratings quoted for capacitors, however, should be regarded as minimum values and capacitors with higher ratings may be used if so desired. In these respects, the home constructor is advised to choose his components very carefully, and complete sets of components or constructional kits purporting to be used with specific Mullard circuits should be examined closely. Mullard Limited only design and publish the circuits: they have no control of material and components sold for use with their circuits.

SMALL COMPONENTS

The smaller components are normally mounted on tagboards, and it is generally more convenient to solder them to the tags before the boards are mounted in the chassis. The small carbon resistors should be laid across two tags and the lead wires bent around the tags in the manner shown in Fig. 1(a). If the tagboard wiring diagrams show that neighbouring tags should be connected together, the lead wires of the

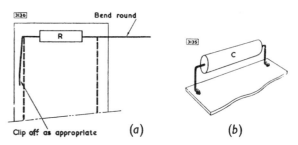

Fig. 1—Mounting of components on tagboards

appropriate components should be cut to lengths sufficient to allow this. Waxed capacitors, silvered-mica capacitors and high-stability resistors will require longer leads so that they will not be overheated when soldered into position. The leads should be bent as shown in Fig. 1(b), care being taken in doing so to avoid making sharp bends very close to the ends of the components.

Components such as the germanium diodes (used in the tape-recorder equipment of Chapters 11 and 12) which are particularly sensitive to temperature, will also require long lead wires. These wires should also be held with a pair of pliers to act as a heat shunt when soldering. The polarity of Mullard germanium

CONSTRUCTION AND ASSEMBLY

diodes is decided by the outer markings: usually the negative pole is the end on which a band is marked.

EARTH CONNECTIONS

In audio equipment, it is advisable to have only one direct connection to the chassis, and this should be made close to the input socket. With a number of connections, eddy currents can be set up in the chassis and hum voltages can be induced in the sensitive sections of the equipment by the magnetic fields associated with these currents. Normally, a common earth lead or busbar is recommended, and the centre spigots of the valveholders often serve as convenient points for securing this lead.

To maintain a low level of hum when two or more units are used together (see Chapter 1), the chassis of the units should be connected by one lead only. If the negative return line is conveyed from one unit to the other by way of the power cable, the connection between chassis should not be duplicated in the a.f. coaxial lead, and one end only of the outer conductor of this lead should be earthed.

Whenever possible, an external connection should be made from the equipment to earth. This lessens hum pick-up and may also help in eliminating any undesirable r.f. signals that can occur in equipment situated near radio or television transmitting stations.

OUTPUT TRANSFORMER

The quality of the output transformer will govern to a great extent the quality of the output from the loudspeaker. A poor component can give rise to a high level of distortion and can cause instability. The most important factors determining the performance of an output transformer are:

1. Primary-winding inductance
2. Leakage inductance
3. Core material and core size
4. Intrinsic capacitance of the coils (especially of the primary)
5. Capacitances between the various coils
6. Power efficiency.

In an actual design, a compromise has to be found between the inductance and capacitance of the primary winding. The inductance must be high enough to minimise the fall in response which occurs at low frequencies. On the other hand, a high inductance necessarily produces a high self-capacitance because of the large number of turns required, and this can result in a large loss in voltage at high frequencies.

The inductance and self-capacitance of the primary winding form a parallel tuned circuit in the output stage, and the resonant frequency of this must be as high as possible. It is usually between 50 and 100kc/s in a good transformer. Furthermore, the capacitance must be low to prevent an increase in the phase shift when feedback is used. A high capacitance can give rise to low-frequency instability in the amplifier. The capacitance can be minimised by careful winding of the primary coil.

If the coupling between the primary and secondary windings of the transformer is not close enough (corresponding to increased leakage inductance) excessive phase shift will be introduced at high frequencies which may lead to parasitic oscillations. Close coupling is often obtained by dividing the primary winding into sections and winding the secondary turns between these sections.

It is advisable to obtain output transformers from manufacturers who build the components for specific circuits. The commercial types listed with the circuits in subsequent chapters should all prove to be satisfactory in the appropriate equipment.

NEGATIVE FEEDBACK

Negative feedback is usually taken from the output transformer to the input stage of a power amplifier to reduce the harmonic distortion in the output signal. The correct polarity of this feedback is essential. Connection to the wrong side of the secondary winding of the output transformer will give positive feedback and will result in oscillation.

The correct connection can best be found by trial and error, and it is advisable to use a low-grade speaker while doing this. With unsuitable phasing (positive feedback) the resultant violent oscillation might damage a delicate speaker-coil suspension. If a low-grade speaker is not available, a momentary completion of the feedback loop should be sufficient to determine the correct phasing.

Should it be desired to experiment with the level of feedback – if r.f. instability occurs, for example – this can be done by altering the value of the resistor in the feedback loop. If the resistance is increased, there is a decrease in the level of feedback. (A corresponding value of shunt capacitance in the feedback loop must be found empirically.) Except in the unlikely event of r.f. instability, little advantage will be gained from changing the values of components recommended in later chapters. Audible improvement is indiscernible when the level of feedback is increased above that indicated for each amplifier.

CONSTRUCTION AND ASSEMBLY

STEREOPHONIC BALANCE CONTROLS

The balance control in stereophonic equipment will often consist of dual-ganged potentiometers connected in the grid circuits of corresponding valves in each channel. One of the potentiometers will be connected normally with its minimum-resistance end earthed and the other will be connected in reverse, with its maximum-resistance end earthed. The ganged potentiometers could be either a pair of linear-law potentiometers or one logarithmic-law and one antilogarithmic-law potentiometer. The characteristics of these components, connected normally and in reverse, are shown in Fig. 2.

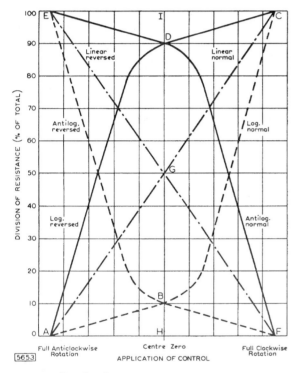

Fig. 2—Potentiometer characteristics

The lines AGC and EGF are the characteristics of normally and inversely connected linear potentiometers respectively. If these are used for the balance control, then the need for the centre-zero arrangement will give an operating point at the intersection G of the two characteristics. The resistance corresponding to IG will appear in series with each grid, and that represented by GH will be between the grid and earth. The signal attenuation caused in each channel by this arrangement will thus be 50%.

The curves ADC and EDF are respectively the characteristics of a logarithmic potentiometer connected in reverse and an antilogarithmic potentiometer connected normally. The operating point with this combination would be the intersection D, and the signal attenuation in the zero position of the control would be only ID/IH. For potentiometers obeying a 10% logarithmic law, the attenuation would thus only be 10%.

POWER SUPPLIES

Allowance has been made in the circuits of subsequent chapters for adequate h.t. power supplies. The requirements of each item of equipment dictate the type of rectifier to be used and the specification for the mains transformer.

A minimum value R_{lim}min. of series resistance is quoted on page 52 for the rectifiers used with the circuits in this book. If the specification of the transformer does not give a total winding resistance equal to or greater than this minimum, then a resistor must be included in each anode circuit. The amount of series resistance R_t contributed at each anode by the transformer is:

$$R_t = R_s + n^2 R_p,$$

where R_s = the resistance of half the secondary winding

R_p = the resistance of the total primary winding

and n = the ratio of the number of turns on half the secondary winding to the number of turns on the whole primary.

The value R of the resistance which must be added to each anode circuit to produce the necessary limiting value R_{lim}min. is therefore given by:

$$R = R_{lim}\text{min.} - R_t.$$

A good mains transformer is essential for high-quality equipment. It may be necessary to screen the transformer to reduce the transfer of hum, and a mains filter may also be required if interference arising from the mains supply is experienced.

The policy of using only one connection to the chassis should also be applied to reservoir and smoothing capacitors. It is not advisable to bolt the negative cans of these capacitors directly to the chassis. The cans should be insulated from the chassis and the negative connection made to the common earth lead.

Low-frequency instability can result from inadequate decoupling of the power supply. If this occurs, an increase in the values of the decoupling components or the addition of extra decoupling components may prove beneficial. The extent of this increase must be determined experimentally.

CHAPTER 5
Twenty-watt Amplifier

The circuit to be described in this chapter is designed to give the highest standard of sound reproduction when used in association with a suitable pre-amplifier, a high-grade pick-up head and a good-quality loudspeaker system.

Two Mullard output pentodes, type EL34, rated at 25W anode dissipation, form the output stage of the circuit. These are connected in a push-pull arrangement with distributed loading, and give a reserve of output power of 20W with a level of harmonic distortion less than 0·05%. The intermediate stage consists of a cathode-coupled, phase-splitting amplifier using the Mullard double triode, type ECC83. This stage is preceded by a high-gain voltage amplifier incorporating the Mullard low-noise pentode, type EF86. Direct coupling is used between the voltage amplifier and phase splitter to minimise low-frequency phase shifts.

The main feedback loop includes the whole circuit, the feedback voltage being derived from the secondary winding of the output transformer and being injected in the cathode circuit of the EF86. The amount of feedback applied around the circuit is 30dB, but in spite of this high level, the stability of the circuit is good and the sensitivity is 220mV for the rated output power. The level of hum and noise is 89dB below the rated 20W.

The rectifier used in the power-supply stage is the Mullard full-wave rectifier, type GZ34. This provides sufficient current for the amplifier (about 145mA) and also for the pre-amplifier and f.m. radio tuner unit (about 40mA) being used with it.

Prototype of Mullard Twenty-watt Amplifier

20W AMPLIFIER

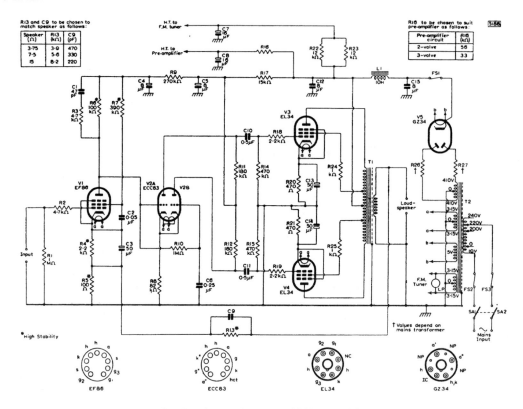

Fig. 1—Circuit diagram of 20W amplifier

CIRCUIT DESCRIPTION

Input Stage

The EF86 input stage of the circuit of Fig. 1 provides high-gain voltage amplification, the stage gain being approximately 120 times. High-stability, cracked-carbon resistors are used in the anode, screen-grid and cathode circuits, and they give an appreciable improvement in the measured level of background noise compared with ordinary carbon resistors.

The stage is coupled directly to the input of the phase splitter. The purpose of this is to minimise low-frequency phase shift in the amplifier and to improve the low-frequency stability when negative feedback is applied. A CR network (C1, R3) connected across the anode load produces an advance in phase and thus improves the high-frequency stability of the amplifier.

Intermediate Stage

The second stage of the circuit uses a Mullard double triode, type ECC83, and fulfils the combined function of phase splitter and driver amplifier. The phase splitter is a cathode-coupled circuit which enables a high degree of balance to be obtained in the push-pull drive signal applied to the output stage.

With the high line voltage available, the required drive voltage for an output power of 20W is obtained with a low level (0.4%) of distortion. The anode load resistors R11 and R12 should be matched to within 5%, R12 having the higher value for optimum operation. Optimum balance is obtained when the effective anode loads differ by 3%. The grid resistors R14 and R15 of the output stage should also be close-tolerance components because they also form part of the anode load of the driver stage. High-frequency balance will be determined largely by the wiring layout because equality of shunt capacitances is required. Low-frequency balance is controlled by the time constant C_6R_{10} in the grid circuits of the triode sections, and the time constant chosen in Fig. 1 will give adequate balance down to very low frequencies.

A disadvantage of the cathode-coupled form of phase splitter is that the effective voltage gain is about half that attainable with one section of the valve used as a normal voltage amplifier. However, as the mutual conductance of the ECC83 is high (100), the effective gain of the cathode-coupled circuit is still about 25 times.

LIST OF COMPONENTS

Resistors

Circuit ref.	Value	Tolerance (±%)	Rating (W)
R1	1 MΩ	20	$\frac{1}{4}$
R2	4·7kΩ	20	$\frac{1}{4}$
R3	4·7kΩ	10	$\frac{1}{4}$
[1]R4	2·2kΩ	10	$\frac{1}{2}$
[1]R5	100 Ω	5	$\frac{1}{2}$
[1]R6	100 kΩ	10	$\frac{1}{2}$
[1]R7	390 kΩ	10	$\frac{1}{2}$
R8	82 kΩ	10	$\frac{1}{4}$
R9	270 kΩ	10	$\frac{1}{2}$
R10	1 MΩ	20	$\frac{1}{4}$
[2]R11	180 kΩ	10	$\frac{1}{2}$
[2]R12	180 kΩ	10	$\frac{1}{2}$
[1]R13 for 3·75Ω speaker	3·9kΩ	5	$\frac{1}{2}$
for 7·5Ω speaker	5·6kΩ	5	$\frac{1}{2}$
for 15Ω speaker	8·2kΩ	5	$\frac{1}{2}$
[3]R14	470 kΩ	10	$\frac{1}{4}$
[3]R15	470 kΩ	10	$\frac{1}{4}$
R16 for 2-valve pre-amp.	56 kΩ	10	1
for 3-valve pre-amp.	33 kΩ	10	1
R17	15 kΩ	20	$\frac{1}{2}$
R18	2·2kΩ	20	$\frac{1}{4}$
R19	2·2kΩ	20	$\frac{1}{4}$
R20	470 Ω	5	3
R21	470 Ω	5	3
R22	12 kΩ	20	6
R23	12 kΩ	20	6
R24	1 kΩ	10	$\frac{1}{2}$
R25	1 kΩ	10	$\frac{1}{2}$
R26 and R27	Values depend on mains transformer		6

1. High stability, cracked carbon
2. Matched to within 5%
3. Preferably matched to within 5%
4. Wire wound

Capacitors

Circuit ref.	Value	Description	Rating (V)
C1	47 pF	silvered mica[5]	
C2	0·05μF	paper	350
C3	50 μF	electrolytic	12
C4	8 μF	electrolytic	450
C5	8 μF	electrolytic	450
C6	0·25μF	paper	350
C7	16 μF	electrolytic	450
C8	8 μF	electrolytic	500
C9 for 3·75Ω speaker	470 pF	silvered mica[6]	
for 7·5Ω speaker	330 pF	silvered mica[6]	
for 15Ω speaker	220 pF	silvered mica[6]	
C10	0·5 μF	paper	350
C11	0·5 μF	paper	350
C12	8 μF	paper	500
C13	50 μF	electrolytic	50
C14	50 μF	electrolytic	50
C15	8 μF	paper	500

5. Tolerance, ±10%
6. Tolerance, ±5%

Valves

Mullard EF86, ECC83, EL34 (two), GZ34

Valveholders

B9A (noval) nylon-loaded, with screening skirt (for EF86) McMurdo, XM9/AU, Skirt 95
B9A (noval) nylon-loaded (for ECC83), McMurdo XM9/AU
B8-O (International octal) (three, for EL34s and GZ34), McMurdo B8/U

Miscellaneous

Mains input plug, 3-way. Bulgin, P340
Mains switch, 2-pole. Bulgin, S300
Mains selector. Clix, CTSP/2
H.T. supply socket (f.m. tuner), 4-way. Elcom, S.04
H.T. supply socket (pre-amplifier), 6-way. Elcom, S.06
Fuseholders (three). Belling Lee, L356

Fuses, 2A (two); 250mA (one)
Lampholder. Bulgin, D180/Red
Pilot lamp. 6.3V, 40mA
Input socket, coaxial. Belling Lee, L.734/S
Output plug, 2-pin. Bulgin, P350
Tagboard (10-way) (two). Bulgin, C114

Output Transformer T1

Primary Impedance:
7kΩ for 20% screen-grid taps
6·6kΩ for 43% screen-grid taps

Mains Transformer T2

Primary: 10–0–200–220–240V
Secondaries: H.T., 410–0–410V, 180mA
3·15–0–3·15V, 4A
3·15–0–3·15V, 2·5A
0–5V, 3A

Smoothing Choke L1

Inductance: 10H at 180mA
Resistance: 200Ω

Commercial Components

Manufacturer	Output Transformer Type No. 20% Taps	Output Transformer Type No. 43% Taps	Mains Transformer Type No.	Choke Type No.
Colne	03070	03069	03068	03071
Elden	486A	486	477	478
Gardners	AS.7034	AS.7034	RS.3175	CS.5142
Gilson	W.O.1342	W.O.866	W.O.775	W.O.1340
			W.O.917	W.O.1341
Hinchley	1532	1377	1441	1528
Parmeko	P2913	P2647	P2646	P463
Partridge	—	P3878	P3877	C10/180
	—	P6878	P6877	—
Savage	—	4B14	4B32-1	—
Wynall	W1900C	W1552C	W1584	W1585

20W AMPLIFIER

Output Stage

The main feature of interest in the output stage is the use of two EL34s with partial screen-grid (or distributed) loading, the screen grids being fed from tappings on the primary winding of the output transformer. As stated in Chapter 3, the best practical operating conditions are achieved with this type of output stage when about 20% of the primary winding is common to the anode and screen-grid circuit.

The anode-to-anode loading of the output stage is 7kΩ and, with a line voltage of 440V at the centre-tap of the primary winding of the output transformer, the combined anode and screen-grid dissipation of the output valves is 28W per valve. With the particular screen-grid-to-anode load ratio used, it has been found that improved linearity is obtained at power levels above 15W when resistors of the order of 1kΩ are inserted in the screen-grid supply circuits. The slight reduction in peak power-handling capacity which results is not significant in practice.

Separate cathode-biasing resistors are used in the output stage to limit the out-of-balance direct current in the primary winding of the output transformer. The use of other balancing arrangements has not been thought necessary although it is likely that some improvement in performance, particularly at low frequencies, would result from the use of d.c. balancing. It is necessary in this type of output stage for the cathodes to be bypassed to earth even if a shared cathode resistor is used. Consequently, a low-frequency time constant in the cathode circuit cannot be eliminated when automatic biasing is used.

Negative Feedback

Negative feedback is taken from the secondary winding of the output transformer to the cathode circuit of the input stage. In spite of the high level of feedback used (30dB), the circuit is completely stable under open-circuit conditions. At least 10dB more feedback (obtainable by reducing the value of R13) would be required to cause high-frequency instability. The most probable form of instability would be oscillation with capacitive loads, but this is most unlikely to occur even with very long loudspeaker leads.

Power Supply

The power supply is conventional and uses a Mullard indirectly-heated, full-wave rectifier, type GZ34, in conjunction with a capacitive input filter. The values of the limiting resistors R26 and R27 will depend on the winding resistances of the mains transformer used.

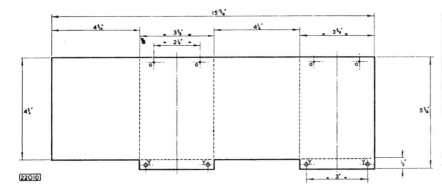

DRILL HOLES IN SCREENING CANS	
Holes	Drill Size
a	49
Y	34
Z	27

Sections of Figs. 2, 3 and 4 should be bent up at 90° at all dotted lines.

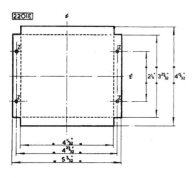 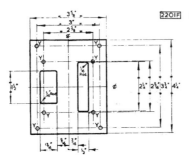

Fig. 2—Screening can for output transformer
  (a) (above) Can
  (b) (far left) Lid
  (c) (near left) Mounting plate

20W AMPLIFIER

Their purpose, when required, is normally one of voltage control only. Where a transformer with a very low winding resistance is used, a secondary voltage rated at 400-0-400V may be found adequate. The rating of the mains transformer is such that about 40mA may be drawn from the h.t. supply to feed a pre-amplifier circuit and f.m. tuner in addition to the normal current required (about 145mA) for the amplifier.

Extra decoupling will be need for these ancillary supplies. The smoothing components R22, R23 and C7 can only be chosen when the type of tuner to be used is known. The values given in Fig. 1 would be suitable for typical current and voltage requirements of approximately 40mA and 200V. The components R16 and C8 depend on the pre-amplifier to be used. The values given in Fig. 1 refer to the 2- and 3-valve pre-amplifiers described in Chapter 9.

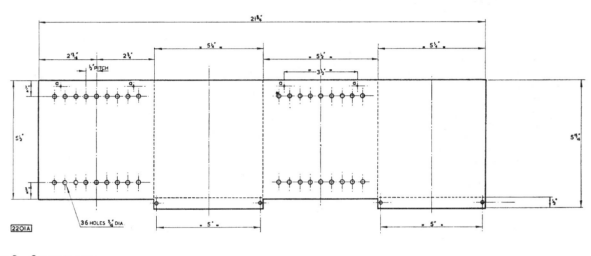

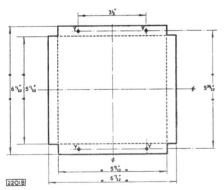
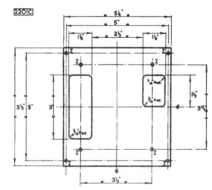

Fig. 3—Screening can for mains transformer

(a) (above) Can

(b) (near right) Lid

(c) (far right) Mounting plate

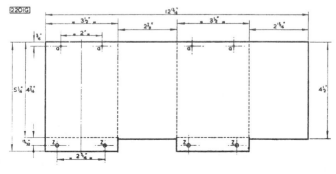
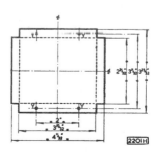

Fig. 4—Screening can for smoothing choke

(a) (near right) Can

(b) (far right) Lid

20W AMPLIFIER

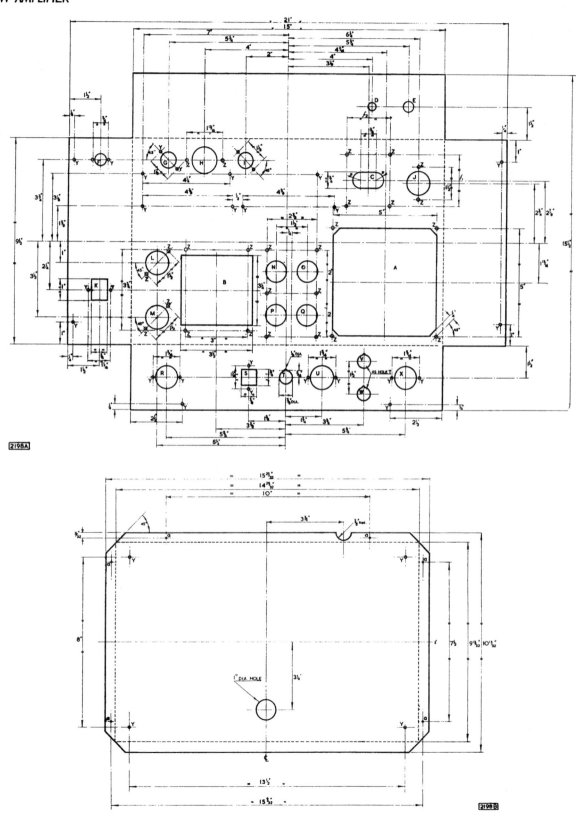

CONSTRUCTION AND ASSEMBLY

The chassis and screening cans for the 20W amplifier are made from ten separate pieces of 16 s.w.g. aluminium sheet. The dimensions (in inches) of these pieces are:

(*a*) Main chassis	$21 \times 15\frac{1}{2}$
(*b*) Base	$15\frac{29}{32} \times 10\frac{13}{32}$
(*c*) Screening can (mains transformer)	$21\frac{9}{16} \times 5\frac{15}{16}$
(*d*) Can lid (mains transformer)	$6\frac{19}{32} \times 6\frac{11}{32}$
(*e*) Mounting plate (mains transformer)	$5\frac{1}{4} \times 5\frac{1}{2}$
(*f*) Screening can (output transformer)	$15\frac{13}{16} \times 5\frac{3}{16}$
(*g*) Can lid (output transformer)	$5\frac{3}{32} \times 4\frac{15}{32}$
(*h*) Mounting plate (output transformer)	$3\frac{5}{8} \times 4\frac{1}{4}$
(*i*) Screening can (smoothing choke)	$12\frac{13}{16} \times 5\frac{1}{16}$
(*j*) Can lid (smoothing choke)	$4\frac{11}{32} \times 3\frac{23}{32}$

Each piece should be marked as shown in the drawings of Figs. 2 to 5, and the holes should be cut as indicated. Where bending is required, it is important for the bends to be made accurately at the lines for the pieces to fit together properly on assembly.

Most of the resistors and small capacitors are mounted on two ten-way tagboards, and they should be connected as indicated in Figs. 6 and 7. The larger components should be fitted to the chassis in the positions indicated in the layout diagram of Fig. 8: this arrangement of the components will ensure good stability.

The wiring between the components is also indicated in Fig. 8. A busbar earth return is indicated, with only a single chassis connection at the input socket. Of the valveholders fitted to the chassis, only the holder for the EF86 needs to be skirted. This holder should also be nylon-loaded.

Fig. 5 (left)—Chassis details (the sections should be bent up at 90° at all dotted lines)

(a) (top) Main chassis. (The dimensions f1 and f2 will depend on the position of the fixing screws on the smoothing choke)

(b) (bottom) Base

KEY TO HOLES IN CHASSIS

Hole	Dimensions	Use	Type No.
A	—	Mains transformer	—
B	—	Output transformer	—
C	—	Smoothing choke	—
D	$\frac{3}{8}$ in. dia.	Pilot lamp. Bulgin	D180/Red
E	$\frac{1}{2}$ in. dia.	Mains switch. Bulgin	S300
F	$\frac{9}{16}$ in. dia.	Input socket, coaxial. Belling Lee	L.734/S
G	$\frac{3}{4}$ in. dia.	B9A nylon-loaded valveholder with screening skirt. McMurdo	XM9/AU, Skirt 95
H	$1\frac{1}{4}$ in. dia.	Electrolytic capacitors	—
I	$\frac{3}{4}$ in. dia.	B9A nylon-loaded valveholder, McMurdo	XM9/AU
J	$1\frac{1}{8}$ in. dia.	B8-O international octal valveholder. McMurdo	B8/U
K	—	H.T. supply socket (for pre-amplifier) 6-way. Elcom	S.06
L	$1\frac{1}{8}$ in. dia.	B8-O international octal valveholder. McMurdo	B8/U
M	$1\frac{1}{8}$ in. dia.	B8-O international octal valveholder McMurdo	B0/U
N, O	1 in. dia.	Paper capacitor	—
P, Q	1 in. dia	Paper capacitor	—
R	$1\frac{1}{16}$ in. dia.	Output socket, 2-pin. Bulgin	P350
S	—	H.T. supply socket (for f.m. tuner) 4-way. Elcom	S.04
T	$\frac{5}{8}$ in. dia.	H.T. fuseholder. Belling Lee	L.356
U	$1\frac{1}{8}$ in dia.	Mains selector. Clix	CTSP/2
V	$\frac{5}{8}$ in. dia.	Mains fuseholder. Belling Lee	L.356
W	$\frac{5}{8}$ in. dia.	Mains fuseholder. Belling Lee	L.356
X	$1\frac{1}{16}$ in. dia.	Mains input socket, 3-pin. Bulgin	P340
Y	Drill No. 34	6B.A. clearance hole	—
Z	Drill No. 27	4B.A. clearance hole	—
a	Drill No. 49		

D.C. CONDITIONS

The d.c. voltages at points in the equipment should be tested with reference to Table 1. The results shown in this table were obtained with an Avometer No. 8.

PERFORMANCE

Distortion

The total harmonic distortion of the prototype amplifier at 400c/s, measured without feedback and with a resistive load, is shown in Fig. 9. The distortion

TABLE 1
D.C. Conditions

Point of Measurement		Voltages (V)	D.C. Range of Avometer* (V)
	C15	465	1000
	C12	440	1000
	C5	410	1000
	C4	160	1000
V3, V4 EL34	Anode	433	1000
	Screen grid	433	1000
	Cathode	32	1000
V2 ECC83	1st and 2nd Anode	325	1000
	1st and 2nd Grid	82·5	1000
	1st and 2nd Cathode	85	1000
V1 EF86	Anode	82·5	1000
	Screen grid	153·5	1000
	Cathode	2·2	25

*Resistance of Avometer:
1000V-range, resistance = 20MΩ
25V-range, resistance = 500kΩ

20W AMPLIFIER

curve towards the overload point is also shown in Fig. 9 when feedback is applied.

At the full rated output, the distortion without feedback is well below 1% and with feedback is below 0·05%. The distortion rises to 0·1% for an output power of 27W. The loop-gain characteristic is such that a level of at least 20dB of feedback is maintained from 15c/s to 25kc/s, and of at least 26dB down to 30c/s.

Measurements of intermodulation products were made in the prototype amplifier using a carrier frequency of 10kc/s and a modulating frequency of 40c/s. The ratio of modulating amplitude to carrier amplitude was 4 : 1. With the combined peak amplitudes of the mixed output at a level equivalent to the peak sine-wave amplitude at an r.m.s. power of 20W, the intermodulation products, expressed in r.m.s. terms, totalled 0·7% of the carrier amplitude. At an equivalent power of 27W, the products totalled 1% of the carrier amplitude. The beat-note distortion between equal-amplitude signals at frequencies of 14 and 15kc/s is 0·25 and 0·3% at equivalent powers of 20 and 27W respectively, and between frequencies of 9 and 10kc/s it is 0·2 and 0·25% at the same equivalent powers. The variations of intermodulation and beat-note distortion with equivalent power are plotted in Fig. 10.

The output/input characteristics of the prototype amplifier are shown in Fig. 9. Excellent linearity is indicated up to output voltages, measured across a 15Ω load, of 20V, which corresponds to an output power of 27W.

Frequency Response

The frequency- and power-response characteristics—that is, the curves for output powers of 1 and 20W respectively—are shown in Fig. 11. From this figure it will be seen that, at an output of 1W, the characteristic is flat ($\pm$1dB) compared with the level at 1kc/s from 2c/s to 100kc/s and the power response characteristic is flat ($\pm$0·5dB) from 30c/s to 20kc/s.

It is important that adequate power-handling capacity is available at the low-frequency end of the audible range and this is determined chiefly by the characteristics of the output transformer. The prototype amplifier is capable of handling powers of at least 20W at frequencies as low as 30c/s without excessive distortion, but for very low frequencies it is desirable that the signal should be attentuated in the associated preamplifier.

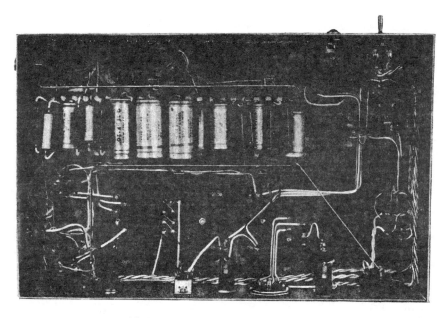

Underside View of Prototype Amplifier

20W AMPLIFIER

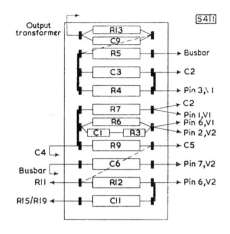

Fig. 6—Tagboard No. 1

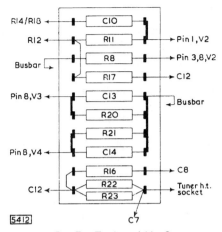

Fig. 7—Tagboard No. 2

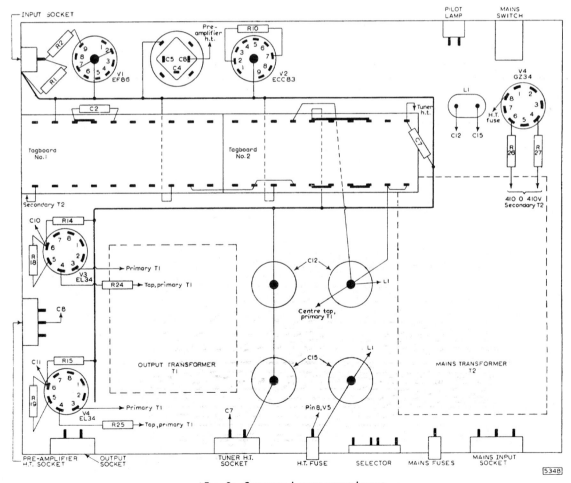

Fig. 8—Suggested component layout

20W AMPLIFIER

Sensitivity

The sensitivity of the amplifier measured at 1kc/s is 6·5mV for an output of 20W when no feedback is applied, and approximately 220mV with feedback, the loop gain being 30dB. The loop gain characteristic of the complete amplifier for the full frequency range is shown in Fig. 11. The sensitivity, with feedback, at the overload point (27W) is approximately 300mV.

The level of background noise and hum in the prototype equipment is 89dB below 20W, measured with a source resistance of 10kΩ. This is equivalent to a signal of about 5·5μV at the input terminals. It is possible to increase the overall sensitivity of the amplifier by 6dB while still maintaining a low background level, a high loop gain and a good margin of stability, but various design requirements of associated pre-amplifier circuits (the need for a high signal-to-noise ratio, for example) render a higher sensitivity a doubtful advantage.

Phase Shift and Transient Response

Emphasis has been laid in the amplifier on a good margin of stability and, consequently, the phase shift is held to a comparatively low level. As shown in Fig. 11, the shift is only 20° at 20kc/s. Excellent response to transient signals is obtained, the rise time of the amplifier being of the order of 5μsec.

Output Impedance

The output stage has a low inherent output impedance, and this is further lowered by the use of negative feedback. It is approximately 0·3Ω with a 15Ω load for an output of 20W at frequencies of 40c/s, 1kc/s and 20kc/s. This gives a damping factor of about 50.

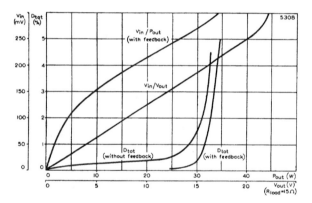

Fig. 9—Harmonic distortion and output/input characteristics

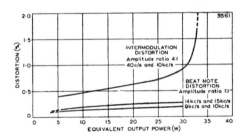

Fig. 10—Intermodulation and 'beat-note' distortion characteristics

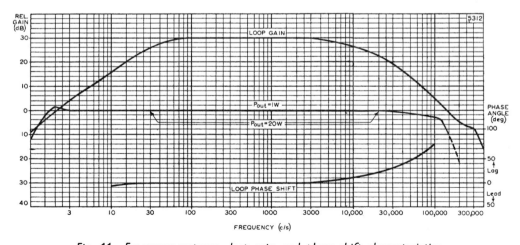

Fig. 11—Frequency-response, loop-gain and phase-shift characteristics

CHAPTER 6

Ten-watt Amplifier

An easily constructed three-stage amplifier with a rated power reserve of 10W is described in this chapter. The circuit is drawn in Fig. 1.

Several versions of the amplifier are possible. The equipment can be constructed to include volume and tone controls, when it is suitable for direct use with high-level signal sources, or it can be made without controls to be used with pre-amplifying equipment. In addition, three arrangements of the output stage are possible: normal loading, low loading and distributed loading.

The push pull output stage of the circuit uses two Mullard output pentodes, type EL84. These yield a rated output power of 10W with a level of harmonic distortion, under normal loading conditions, of less than 0·3%.

The intermediate stage consists of a cathode-coupled phase-splitting amplifier using the Mullard double triode, type ECC83, which is preceded by a high-gain voltage amplifier using the Mullard low-noise pentode, type EF86. The intermediate and input stages are coupled directly to each other so that low-frequency phase shifts are minimised.

The optional treble and bass tone controls are shown in Fig. 1. RV28 provides continuously variable treble control from $+10$ to -10dB at 10kc/s and RV30 provides bass control from $+11$ to -5dB at 20c/s.

The level of feedback taken from the output transformer to the cathode circuit of the input stage is 26dB. The basic sensitivity of the circuit with this amount of feedback is 40mV for the rated output power, and the sensitivity when the tone controls are included is 600mV. The level of noise and hum is at least 75dB below 10W.

The rectifier used in the power-supply stage is the Mullard full-wave rectifier type EZ81.

Prototype of Ten-watt Amplifier with Controls

10W AMPLIFIER

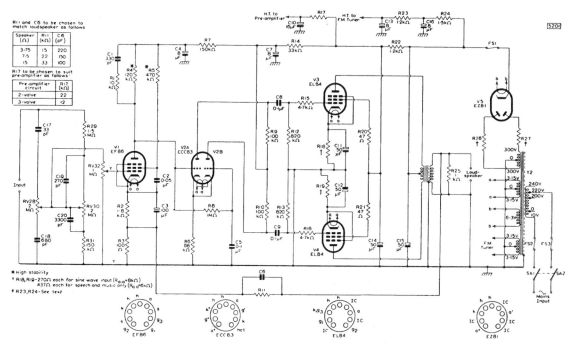

Fig. 1—Circuit diagram of 10W amplifier. (For version without controls, the signal is taken directly to the control grid of V1, and RV32 is replaced by a fixed resistance of 1MΩ.)

LIST OF COMPONENTS

Output Transformer

Primary Impedance:
- Low loading,   6kΩ
- Normal loading,   8kΩ
- Distributed loading, 6·6kΩ for 20% screen-grid taps
-                         8kΩ for 43% screen-grid taps

Commercial Components

Manufacturer	Type No.			
	Low Loading	Normal Loading	Distributed Loading 20%	43%
Colne	03096	03095	03098	03097
Elden	—	—	546A	546
Elstone	OT/6	OT/8	—	OT/ML
Gardners	*AS.7015		AS.7012	
Gilson	W.O.696B	W.O.696A	W.O.710	W.O.892
Haddon	—	—		KD245
				KD246
Hinchley	1533	1376	1533	1376
Parmeko		*P2629	P2643	P2642
Partridge		*P3667	—	P4131
				P4014
Savage	—	—	—	4A93
Wynall	W1899	W1898	W1897	W1494

*8kΩ primary tapped at 6kΩ

Mains Transformer T2

The choice of mains transformer depends on various voltage and current requirements of the circuit to be constructed, particularly on whether normal or low-loading conditions are to be used and if additional units such as an f.m. tuner unit or a pre-amplifier are to be supplied.

Commercial Components

Manufacturer	Type No.	Ratings H.T.	Heaters
Colne	03094	70mA 6·3V, 2·3A, ct.	6·3V, 1A
Elden	396	120mA 6·3V, 3A, ct.	5V, 2A
Elstone	MT5/10	100mA, 6·3V, 2A, ct.	0–5–6·3V, 2A
	MT/MU	120mA 2×6·3V, 2A, ct.	6·3V, 1A
Gardners	RS.3110	150mA 2×6·3V, 2A, ct.	6·3V, 1A
	RS.3103	80mA 6·3V, 2A, ct.	6·3V, 1A
Gilson	W.O.741A-B	150mA 6·3V, 5A, ct.	5V, 2A
	W.O.823	140mA 2×6·3V, 2A, ct.	6·3V, 1A
Haddon	KD241	135mA 2×6·3V, 2A, ct.	6·3V, 1A
	KD242	De-luxe potted version of KD241	
	KD243	110mA 6·3V, 2·5A, ct.	0–5–6·3V, 2A
	KD244	De-luxe potted version of KD243	
Hinchley	1440	120mA 2×6·3V, 2A, ct.	6·3V, 1A
Parmeko	P2631	60mA 6·3V, 2A, ct.	6·3V, 1A
	P2630	100mA 6·3V, 2A, ct.	5V, 2A
Partridge	P4013	135mA 2×6·3V, 2A, ct.	6·3V, 1A
	H300/110	110mA 6·3V, 2·5A, ct.	0–5–6·3V, 1A
Savage	4A97-1	120mA 6·3V, 2·5A, ct.	5V, 2A
Wynall	W1497	125mA 2×6·3V, 2A, ct.	0–5–6·3V, 1A

10W AMPLIFIER

LIST OF COMPONENTS (*continued*)

Resistors

Circuit ref.	Value	Tolerance (±%)	Rating (W)
R1	10 kΩ	10	¼
R2	1·8 kΩ	10	¼
R3	100 Ω	5	¼
[1]R4	120 kΩ	10	¼
[1]R5	470 kΩ	10	½
R6	68 kΩ	10	1
R7	150 kΩ	10	¼
R8	1 MΩ	10	¼
[2]R9	100 kΩ	10	½
[2]R10	100 kΩ	10	½
R11			
for 3·75Ω speaker	15 kΩ	5	½
for 7·5Ω speaker	22 kΩ	5	½
for 15Ω speaker	33 kΩ	5	½
[3]R12	820 kΩ	10	¼
[3]R13	820 kΩ	10	¼
R14	33 kΩ	10	¼
R15	4·7 kΩ	20	¼
R16	4·7 kΩ	20	¼
R17			
for 2-valve pre-amp	22 kΩ	10	½
for 3-valve pre-amp	12 kΩ	10	½
R18			
for normal loading	270 Ω	5	3
for low loading	390+47 Ω	5	3
R19			
for normal loading	270 Ω	5	3
for low loading	390+47Ω	5	3
R20	47 Ω	20	¼
R21	47 Ω	20	¼
R22	1·2 kΩ	10	1
R23	1·2 kΩ	20	3
R24	1·5 kΩ	20	3
R25	1 kΩ	20	½
R26, R27 values depend on mains transformer		20	1
[4]R33	1 MΩ	20	¼
(For control network only)			
RV28	2 MΩ logarithmic potentiometer		
R29	1·5 MΩ	10	¼
RV30	2 MΩ logarithmic potentiometer		
R31	150 kΩ	10	¼
RV32	1 MΩ logarithmic potentiometer		

1. High stability, cracked carbon
2. Matched to within 5%
3. Preferably matched to within 5%
4. In place of volume control RV32 in control-less amplifier

Valves

Mullard EF86, ECC83, EL84 (two), EZ81 (EZ80)

Valveholders

B9A (noval) nylon-loaded with screening skirt (for EF86) McMurdo, XM9/AU, Skirt 95

B9A (noval) (four for ECC83, EL84s and EZ81) McMurdo, BM9/U

Capacitors

Circuit ref.	Value	Description	Rating (V)
C1	330 pF	silvered mica[5]	
C2	0·05 µF	paper	350
C3	100 µF	electrolytic	12
C4	8 µF	electrolytic	350
C5	0·1 µF	paper	350
C6			
for 3·75Ω speaker	220 pF	silvered mica[6]	
for 7·5Ω speaker	150 pF	silvered mica[6]	
for 15Ω speaker	100 pF	silvered mica[6]	
C7	8 µF	electrolytic	350
C8	0·1 µF	paper	350
C9	0·1 µF	paper	350
C10	16 µF	electrolytic	350
C11	50 µF	electrolytic	25
C12	50 µF	electrolytic	25
C13	8 µF	electrolytic	350
C14, C15	50+50 µF	double electrolytic	350
C16	8 µF	electrolytic	350
(For control network only)			
C17	33 pF	silvered mica[5]	
C18	680 pF	silvered mica[5]	
C19	270 pF	silvered mica[5]	
C20	3300 pF	silvered mica[5]	

5. Tolerance: ±10%
6. Tolerance: ±5%

Miscellaneous
(for circuit with controls)

Mains input plug, 3-pin. Bulgin, P340
Mains switch, 2-pole. Bulgin, S300
Mains selector. Clix, CTSP/2
H.T. supply socket (f.m. tuner), 4-way. Elcom, S.04
H.T. supply socket (pre-amplifier), 6-way. Elcom, S.06
Fuseholder, Minifuse (three). Belling Lee, L.575
Fuse, 2·5A (two)
Fuse, 150mA
Lampholder. Bulgin, D180/Red
Pilot lamp, 6·3V, 40mA
Input socket, coaxial. Belling Lee, L.734/S
Output plug. 2-pin. Bulgin, P350
Tagboard (10-way). Bulgin, C114
Tagboard (5-way). Bulgin, C109

Miscellaneous
(for circuit without controls)

Mains input plug, 3-pin miniature. Bulgin, P429
Mains voltage selector, fused. Clix, VSP393/2, P62/1
H.T. supply plug (two), 6-pin miniature. Bulgin P427
Fuseholder, Minifuse. Belling Lee, L.575
Fuse, 250mA
Input socket, coaxial. Belling Lee, L.734/S
Output socket, 2-pin. Painton, 313263
Tagboard (10-way) (two). Bulgin, C125; Denco

10W AMPLIFIER

CIRCUIT DESCRIPTION

Input Stage

The first stage of amplification is provided by the EF86 in a circuit giving a voltage gain of about 150 times. The negative feedback voltage from the secondary winding of the output transformer is introduced across the 100Ω resistor R3 in the cathode circuit. Because of the high gain in this stage, any excessive noise will lower the quality of the output from the speaker considerably and so high-stability, cracked-carbon resistors are used for the anode and screen-grid circuits. The EF86 is coupled directly to the phase splitter to reduce the phase shift at low frequencies. A CR network (C1, R1) shunting the anode load produces an advance in phase which increases the stability of the amplifier at high frequencies.

Intermediate Stage

The double triode type ECC83 is operated as a cathode-coupled phase splitter to provide a push-pull drive voltage for the output stage. The use of the cathode-coupled arrangement gives low distortion and facilitates direct coupling with the input stage. The voltage gain obtained with a cathode-coupled circuit is approximately half that obtainable from each valve section operated as a normal voltage amplifier. Nevertheless, it is sufficient because the amplification factor of the ECC83 is 100.

It is necessary in a cathode-coupled phase splitter for the anode load of the earthed section to be slightly higher than that of the first section if perfect balance is to be obtained. Thus R10 should ideally be slightly higher than R9, by a percentage which depends on the amplification factor of the valve. Because of the high amplification factor of the ECC83, nominally equal resistors, matched to within 5%, cannot give rise to more than 2% lack of balance.

At low frequencies, the presence of C5 and R8 in the grid circuit of the second triode section produces both phase and amplitude unbalance. The frequency at which the lack of balance becomes significant depends on the time constant $C_5 R_8$, which should be as long as possible to provide maximum decoupling within the pass band of the amplifier. Adequate balance can easily be maintained in practice and the component values used in this circuit give it down to low audible frequencies.

The anode voltage of the EF86 determines the operating conditions of the phase splitter and, provided the values of the associated components are within the stated tolerances, the stage will operate correctly. If the anode voltage of the EF86 is too high, the bias on the phase splitter will be too low and grid current distortion, resulting from overdrive on peak signals, will occur. If the anode voltage is too low, the phase splitter bias will be too high, and the stage will operate away from the linear position of the valve characteristics with a resulting increase in distortion.

Output Stage

The output stage consists of two output pentodes, type EL84, in a cathode-biased, push-pull circuit. The anode supply voltages are taken from the reservoir capacitor C15, while the supply voltages for the screen grids and for the rest of the amplifier are taken by way of R22 and C14.

Compensation for the slightly different currents drawn by any two EL84s is achieved by using the separate cathode resistors R18 and R19. Matching of the output valves is therefore unnecessary.

The stopper resistors R15 and R16 in the control-grid leads and R20 and R21 in the screen-grid circuits have been included as a normal measure to prevent parasitic oscillations. These resistors should not be bypassed, and a close connection should be made to the tags on the valveholders.

A resistor R25 having a value of about 1kΩ may be placed across the output terminals to prevent instability from occurring when the loudspeaker is disconnected.

Operating Conditions—Normal, Low and Distributed Loading in Output Stage

Three ways of operating the output stage are possible. These use the conditions for normal loading, low loading and distributed loading.

Normal- or low-loading conditions are determined by the choice of value for the equal cathode resistors R18 and R19 in Fig. 1. For normal loading, the value of each resistor should be 270Ω, and for low loading it should be 437Ω (that is, 390Ω+47Ω). The conditions for distributed loading are achieved by including part of the anode load (that is, part of the primary winding of the output transformer) in the screen-grid circuits of the valves.

Normal loading is used when the amplifier is to be tested up to a full output power with a sine-wave input. Low loading is used only with speech and music signals when it gives reduced distortion. Distributed loading conditions are used to achieve a compromise between the performance of the EL84s when connected as triodes and as pentodes. A sufficient reserve of power is maintained, yet distortion is much less than with the usual pentode connections.

10W AMPLIFIER

Prototype of Ten-watt Amplifier without Controls

Normal and Low Loading

The normal loading conditions correspond very closely to the recommendations for Class AB operation of the EL84. The difference is that in Fig. 1 separate cathode resistors R18 and R19 (each equal to 270Ω) are used instead of a shared resistor of 130Ω. The anode-to-anode load resistance for these conditions is 8kΩ and the anode current for zero input signal is 2×36mA.

The low-loading adjustment is so called because, with the increased values of cathode resistance, the anode-to-anode load resistance is reduced to 6kΩ and the quiescent current to 2×24mA. Although the output stage is connected for cathode-bias operation, operation with speech or music input signals approximates very closely to fixed-bias conditions. A comparison has been made in Chapter 3 between the distortion in the output stage when speech or music signals are used with normal and low loading conditions and this comparison is shown in Fig. 1 on page 19.*

Curves 1 and 3 in that figure have been plotted for fixed bias conditions. Curve 1 shows that distortion increases seriously above 10W when normal loading is used. Curve 3 is for low loading and illustrates the lower level of distortion to be achieved with this adjustment. Curve 2 has been plotted for cathode bias and therefore refers to a sine-wave input. The curve is for normal loading because this is required for sine-wave testing up to the maximum output.

With speech or music inputs, the maximum rated power of the amplifier is only required in the output stage for a small part of the time, the average requirement being comparatively low. However, a large power reserve must be available to cater for the widely differing sound levels which can occur.

The h.t. current consumption is smaller when the output stage is adjusted for low loading. Consequently, the standing dissipation in this stage is reduced from 11W at each anode with normal loading to 7·5W at each anode. The valves are thus being run well below the maximum anode-dissipation rating of 12W. The h.t. current requirement is reduced with low loading so that the mains transformer rating can be lower if the amplifier is to be used permanently with the low-loading adjustment.

Larger peak currents are produced in the output stage under low-loading conditions than with normal Class AB operation. With speech or music inputs, these peaks are of short duration. They are supplied by the reservoir capacitor C15 which is large enough to provide satisfactory smoothing. When the amplifier is at the point of overload on a peak signal, the momentary fall in line voltage from a nominal value of 320V should not be more than 2V.

As the current in the output stage increases, the bias voltages across the cathode resistors also increase at a rate determined by the time constants of the bias network. It is unlikely that this increase will exceed 1V. The working conditions of the output stage are such that the output valves are driven back into a region where lower distortion is obtained. As a result of this change in bias, a variation in gain will occur, but the distortion introduced by gain variation is held to a low level by the large amount of negative feedback. The total effect, therefore, is to improve the performance at peaks of output power.

---

*The distortion curves of Fig. 1, Chapter 3 are for the output stage only. Distortion in the complete amplifier is discussed below.

It will be realised that, when the amplifier is being used for its primary purpose as a means of home entertainment, the low-loading adjustment is to be preferred to normal loading. However, sine-wave testing up to the full output power should not be undertaken with the low-loading arrangement. Such testing should only be undertaken with normal loading. Measurements of frequency response may be made with low-level sinusoidal inputs provided the output power does not exceed 1 to 1·5W and square-wave testing may also be undertaken if the signal is again restricted in the same way as the sine-wave input. It is not felt that these restrictions will be of very great significance because few home constructors will have the equipment needed for sine-wave and square-wave testing.

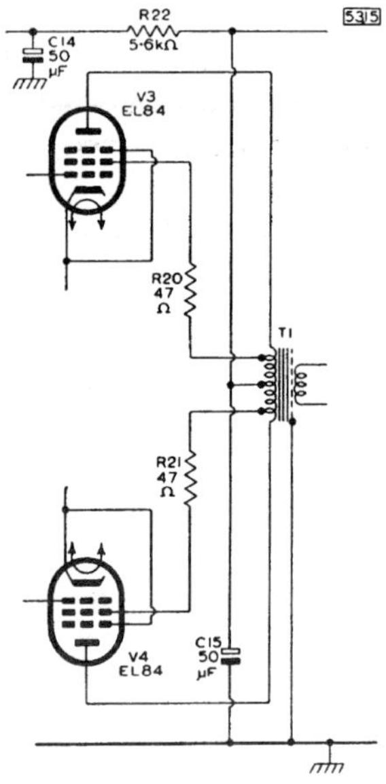

Fig. 2—Arrangement of output stage for distributed loading

Distributed Load

The arrangement of the output stage for operation under conditions of distributed load is shown in Fig. 2. The screen-grid resistors R20 and R21 are disconnected from C14 in Fig. 1 and are connected to the tappings provided on the primary winding of the output transformer. The centre tap of the primary winding is connected, as before, to the reservoir capacitor C15. The dropper resistor R22 in the h.t. line no longer carries the screen-grid current and so is increased from 1·2kΩ to 5·6kΩ to maintain the same d.c. conditions in the first two stages of the amplifier.

The best results are obtained with each half of the primary winding tapped so that about 20% of the turns are common to the anode and screen-grid circuits, when the anode-to-anode loading should be 6·6kΩ.

With distributed loading, total distortion is reduced very much compared with normal loading. The rated power reserve of 10W is maintained although overloading now occurs at 11W instead of 14W. However, the rate at which distortion increases with output power after overloading is less with distributed than with normal loading. The frequency response of the amplifier is the same for either method of loading, although the stability is improved with distributed loading because the lower level of distortion is obtained with reduced loop gain.

Negative Feedback

The level of negative voltage feedback taken from the secondary winding of the output transformer to the cathode circuit of the input valve is 26dB. The output impedance with feedback is 0·9Ω measured at the 15Ω output terminals. This gives a satisfactory damping factor of about 17 (that is, 15/0·9).

Power Supply

The power-supply stage shown in Fig. 1 uses the Mullard indirectly-heated, full-wave rectifier, type EZ81. Adequate smoothing is achieved using the

Fig. 3 (right)—Chassis details for amplifier with volume and tone controls (the sections should be bent up at 90° at all dotted lines)

(a) Main chassis

(b) Base

(c) Screen

KEY TO HOLES IN CHASSIS (WITH CONTROLS)

Hole	Dimension	Use	Type No.
A	3/8 in. dia.	2MΩ potentiometer (logarithmic) ..	—
B	3/8 in. dia.	2MΩ potentiometer (logarithmic) ..	—
C	3/8 in. dia.	1MΩ potentiometer (logarithmic) ..	—
D	3/8 in. dia.	Pilot lamp. Bulgin	D180/Red
E	1/2 in. dia.	Mains switch. Bulgin	S300
F	3/4 in. dia.	B9A nylon-loaded valveholder with screening skirt. McMurdo	XM9/AU Skirt 95
G	3/4 in. dia.	B9A valveholder. McMurdo	BM9/U
H	1 1/2 in. dia.	Electrolytic capacitor	—
I	9/16 in. dia.	Input socket, coaxial. Belling Lee.. ..	L.734/S
J	3/4 in. dia.	B9A valveholder. McMurdo	BM9/U
K	3/4 in. dia,	B9A valveholder. McMurdo	BM9/U
L	3/4 in. dia.	B9A valveholder. McMurdo	BM9/U
M	1 1/16 in. dia.	Output plug, 2-pin. Bulgin	P350
N	1 in. × 3/4 in.	H.T. supply socket, 6-pin. Elcom	S.06
O	3/4 in. × 3/4 in.	H.T. supply socket, 4-pin. Elcom	S.04
P	1 1/8 in. dia.	Mains selector. Clix	CTSP/2
Q	7/16 in. dia.	Mains fuseholder, Minifuse. Belling Lee ..	L.575
R	7/16 in. dia.	Mains fuseholder, Minifuse. Belling Lee ..	L.575
S	1 1/16 in. dia.	Mains input plug, 3-pin. Bulgin	P.340
T	7/16 in. dia.	H.T. fuseholder, Minifuse. Belling Lee ..	L.575
Y	Drill No. 34	6 B.A. clearance hole	—

10W AMPLIFIER

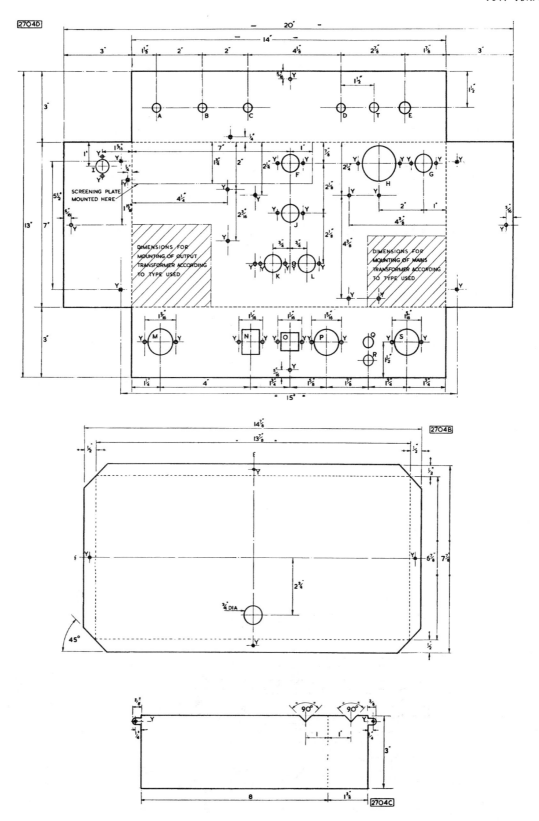

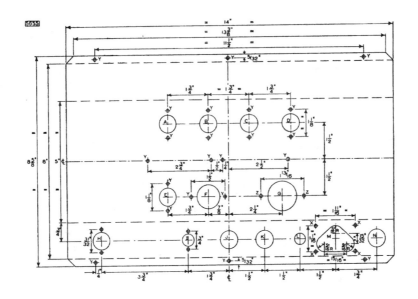

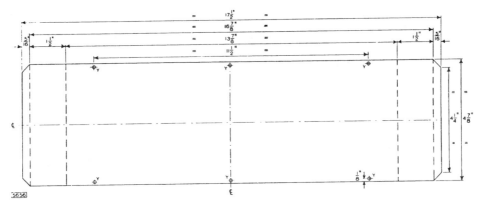

Fig. 4—Chassis details for amplifier without volume and tone controls (the sections should be bent up at 90° at the dotted lines)
(a) Main chassis
(b) Base

KEY TO HOLES IN CHASSIS (WITHOUT CONTROLS)

Hole	Dimension	Use	Type No.
A	$\frac{3}{4}$ in. dia.	B9A valveholder. McMurdo	BM9/U
B	$\frac{3}{4}$ in. dia.	B9A valveholder. McMurdo	BM9/U
C	$\frac{3}{4}$ in. dia.	B9A valveholder. McMurdo	BM9/U
D	$\frac{3}{4}$ in. dia.	B9A valveholder. McMurdo	BM9/U
E	$\frac{3}{4}$ in. dia.	B9A nylon-loaded valveholder with screening skirt. McMurdo	XM9/AU, skirt 95
F	1 in. dia.	Electrolytic capacitor	—
G	$1\frac{1}{4}$ in. dia.	Electrolytic capacitor	—
H	$\frac{11}{16}$ in. dia.	Output socket, 2-pin. Painton	313263
I	$\frac{1}{2}$ in. dia.	Input socket, coaxial. Belling Lee ..	L.734/S
J	$\frac{3}{4}$ in. dia.	H.T. supply plug, 6-pin miniature. Bulgin	P427
K	$\frac{3}{4}$ in. dia.	H.T. supply plug, 6-pin miniature. Bulgin	P427
L	$\frac{7}{16}$ in. dia.	Fuseholder, Minifuse. Belling Lee ..	L.575
M	—	Mains voltage selector, fused. Clix ..	VSP393/2, P62/1
N	$\frac{3}{4}$ in. dia.	Mains input plug, 3-pin miniature. Bulgin	P429
X	3 mm dia.		
Y	Drill No. 34	6 B.A. clearance hole	—

resistor R22 and the large capacitor C15 (50μF) so that the expense of smoothing chokes is avoided. The current supplied by the EZ81 is sufficient for any arrangement of the output stage and also allows the use of a radio tuner unit. If the low-loading version of the output stage is adopted and it is not intended to use the amplifier with an f.m. tuner unit, then the Mullard rectifier, type EZ80, can be used in place of the EZ81, but it must be emphasised that currents greater than 90mA must not be taken from the EZ80.

Limiting resistors must be included in the anode circuits of the rectifier, and their values will depend on the type of mains transformer used. The total resistance – made up of the added resistance and the

10W AMPLIFIER

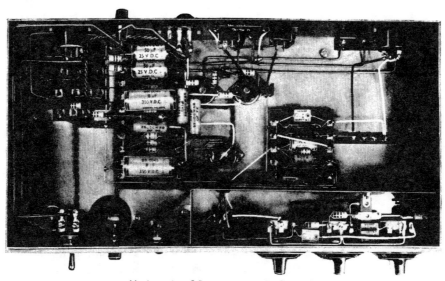

Underside of Prototype with Controls

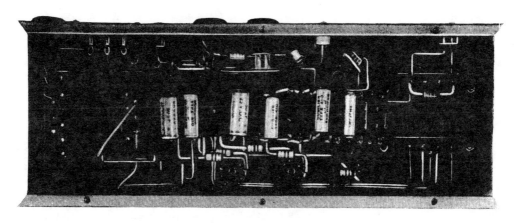

Underside of Prototype without Controls

effective resistance of the mains transformer – must be at least 200Ω per anode for the EZ81 used in this particular circuit (215Ω for the EZ80). For most 300-0-300V mains transformers, the effective resistance is about 130Ω so the value of R26 and R27 (in the 20% range of preferred values) must be at least 85Ω each (100Ω for the EZ80). With the power supply giving an h.t. current of 80mA, the voltage across C15 should then be near the specified 320V. The voltage across C15 must not exceed 320V at either 80mA with normal loading, ($R_{a-a}=8kΩ$) or 60mA with low loading ($R_{a-a}=6kΩ$).

When an f.m. radio tuner unit is used with the amplifier, the additional h.t. current drawn will not normally exceed 35 to 40mA and the voltage across C15 will then drop to approximately 295V. This additional current should be taken from C15 by way of suitable dropping resistors R23 and R24 and should have adequate decoupling (provided by C13 and C16). In these circumstances, the overload point of the amplifier is reduced from 13 or 14W to approximately 12W.

The h.t. supply for pre-amplifiers should be taken from C14. The values of the smoothing resistor R17 given in the table in Fig. 1 are suitable for use with the 2- and 3-valve pre-amplifiers of Chapter 9.

10W AMPLIFIER

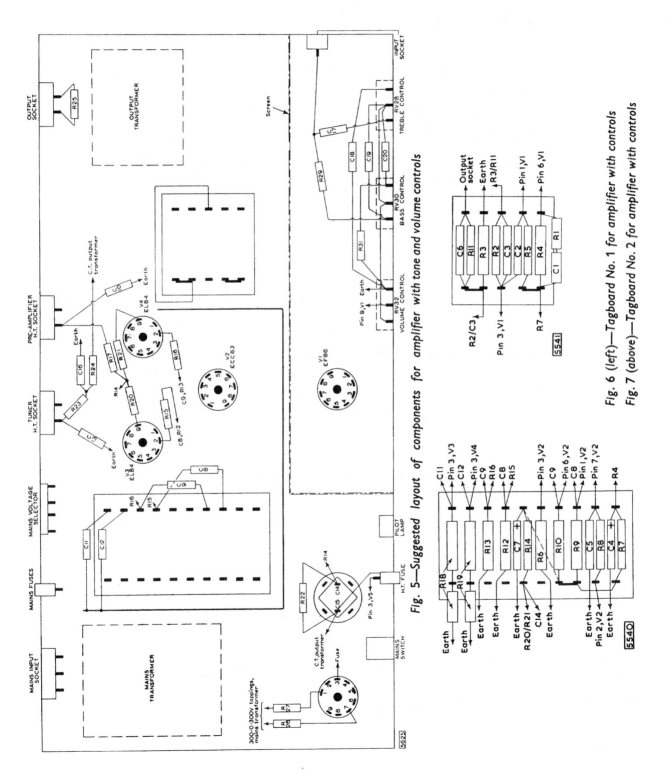

Fig. 5—Suggested layout of components for amplifier with tone and volume controls

Fig. 6 (left)—Tagboard No. 1 for amplifier with controls
Fig. 7 (above)—Tagboard No. 2 for amplifier with controls

10W AMPLIFIER

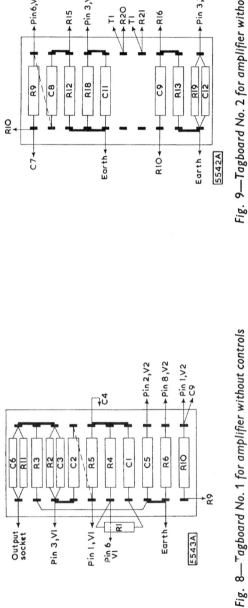

Fig. 8—Tagboard No. 1 for amplifier without controls

Fig. 9—Tagboard No. 2 for amplifier without controls

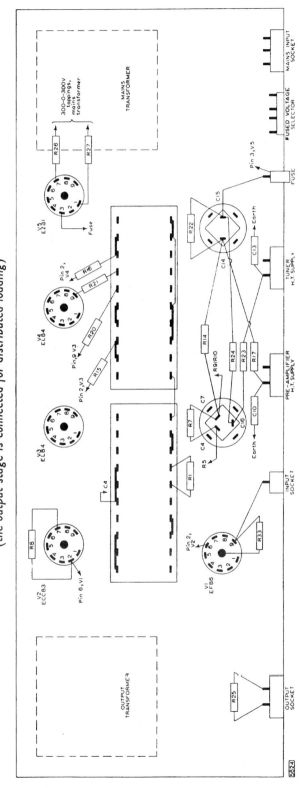

Fig. 10—Suggested layout of components for amplifier without tone and volume controls (the output stage is connected for distributed loading)

10W AMPLIFIER

CONSTRUCTION AND ASSEMBLY
The chassis for the 10W amplifier in which tone and volume controls are incorporated is made from three separate pieces of 16 s.w.g. aluminium sheet. Two separate pieces of 16 s.w.g. aluminium sheet are required for the version without the controls. The dimensions (in inches) of these pieces are:

A – For version with controls
- (a) Main chassis 20×13
- (b) Base $14\frac{7}{8} \times 7\frac{7}{8}$
- (c) Screen 8×3

B – For version without controls
- (a) Main chassis $14 \times 8\frac{5}{8}$
- (b) Base $17\frac{1}{2} \times 4\frac{7}{8}$

The pieces should be marked as shown in either Fig. 3 (with controls) or Fig. 4 (without controls) and the holes should be cut as indicated. Where bending is required, it is important for the bends to be made accurately at the lines for the pieces to fit together properly on assembly.

Two sets of assembly drawings are included in this chapter. One set is for the version of the amplifier which includes the volume and tone controls and the other set is for the control-less version of the amplifiers. The versions differ considerably in layout, but the two arrangements will give good stability. The layout of the control-less version has been dictated to some extent by the requirements of stereophonic equipment. The narrow chassis with the sockets arranged on one side will enable two amplifiers to be used back-to-back with considerable economy of space. A symmetrical arrangement of two amplifiers can be achieved if one of the units is built as a mirror image of the other.

In the version of the amplifier having controls, the majority of resistors and small capacitors are assembled on one ten-way tagboard and one five-way board. These are drawn in Figs. 6 and 7 respectively. The positions of these boards and the other components is shown in Fig. 5, and the wiring between the components is also shown in this figure. A busbar earth return is used and only one connection is made between this and the chassis. This connection is made, via the volume-control potentiometer, at the input socket. The only valveholder which needs to be skirted is that for the EF86, and it should also be nylon-loaded.

Most of the components of the control-less version of the amplifier are assembled on two ten-way tagboards and these are drawn in Figs. 8 and 9. The general layout of the components in the chassis is shown in Fig. 10. As before, a busbar earth return is used and again there is only one connection, at the input socket, between it and the chassis. Only the valveholder for the EF86 needs to be skirted and it should also be nylon-loaded.

D.C. CONDITIONS
The d.c. voltages at points in the equipment should be tested with reference to Table 1. The results shown in this table were obtained using an Avometer No. 8.

PERFORMANCE
Distortion
The total harmonic distortion of the prototype amplifier with normal loading and with feedback was measured at 400c/s. The variation with output power is shown in Fig. 11. The relationship between input voltage and output power is also drawn in Fig. 11. At the full rated output power, the distortion is less than 0·3% and at about 14W it rises to 1%.

Measurements of intermodulation distortion were made in the prototype amplifier using carrier and

TABLE 1
D.C. Conditions

	Point of Measurement	Voltages (V)	D.C. Range of Avometer* (V)
	†C15	320	1000
	C14	305	1000
	C7	235	1000
	C4	100	1000
V3, V4 EL84	Anode	310	1000
	Screen grid	305	1000
	Cathode	12	100
V2 ECC83	1st and 2nd Anodes	185	1000
	1st and 2nd Grids	70	1000
	1st and 2nd Cathodes	71·5	1000
V1 EF86	Anode	70	1000
	Screen grid	85	1000
	Cathode	1·5	25

*Resistance of Avometer:
1000V-range, resistance=20MΩ
100V-range, resistance=2MΩ
25V-range, resistance=500kΩ

†Alternating voltage at C15 is 4V for normal loading and 2·5V for low loading.

10W AMPLIFIER

modulating frequencies of (a) 10kc/s and 40c/s, and (b) 7kc/s and 70c/s, the ratios of the amplitudes being 1:4. The values are plotted in Fig. 12 against equivalent output power. For a stated equivalent output power, the peak value of the combined inputs is equal to the peak value of a single sine-wave input which would produce that power. For frequencies of 10kc/s and 40c/s, the intermodulation distortion (expressed in r.m.s. terms as the percentage of intermodulation products relative to the amplitude of the higher frequency) is better than 1·0% at 10W equivalent output power.

The magnitude of 'beat-note' distortion has been measured with frequencies of (a) 9 and 10kc/s, and (b) 14 and 15kc/s, the amplitudes of the input signals being equal. The difference component present in the output, expressed as a percentage of either signal, is also plotted against equivalent output power in Fig. 12.

Sensitivity
The basic sensitivity of the amplifier (that is, the sensitivity at the control grid of the EF86) is 2mV without feedback and 40mV with feedback. The sensitivity at the input terminals in Fig. 1, thus including the controls, is 600mV with feedback. The basic sensitivity at the overload point (14W output) measured with feedback is 50mV.

Hum and Noise
The level of background noise and hum in the prototype equipment is at least 75dB below the rated output of 10W.

Phase Shift
The phase shift in the amplifier is low. As shown in Fig. 13, the shift is 20° at 20c/s and at 10kc/s.

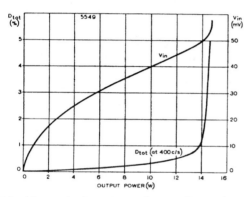

Fig. 11—Harmonic distortion and output/input characteristics

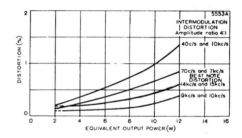

Fig. 12—Intermodulation and 'beat-note' distortion characteristics

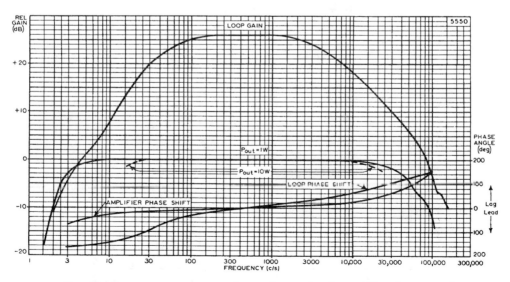

Fig. 13—Frequency response, loop-gain and phase-shift characteristics

10W AMPLIFIER

Frequency Response
The frequency-response characteristic for an output power of 1W and the power-response characteristic for 10W output power under normal loading conditions are shown in Fig. 13. It will be seen that the frequency response is level (to within 1dB) from 5c/s to 20kc/s and the power response is level from 20c/s to 15kc/s.

The loop gain characteristic is also plotted in Fig. 13. This curve shows that 26dB of feedback are available at 1kc/s.

Output Impedance
The output impedance of the amplifier is approximately 0.9Ω for a 15Ω load. This gives an adequate damping factor of about 17.

Performance with Normal and Distributed Loading
Performance figures are given in Table 2 comparing the amplifier when a conventional pentode push-pull stage is used with the unit when distributed loading is used.

TABLE 2
Performance with Normal and Distributed Loading

	Normal Loading	Distributed Loading	
Rated output power	10	10	W
Overload point	14 (approx.)	11 (approx.)	W
Basic sensitivity	40	40	mV
Harmonic distortion	0.3	0.1 (approx.)	%
Intermodulation distortion			
(i) 10kc/s and 40c/s	1.0	0.5	%
(ii) 7kc/s and 70c/s	0.6	0.4	%
Beat-note distortion			
(i) 9 and 10kc/s	0.25 (approx.)	0.25	%
(ii) 14 and 15kc/s	0.4	0.33	%
Loop gain (at 1kc/s)	26	20.5	dB

LIMITING RESISTORS FOR MULLARD RECTIFIERS
(relevant to circuits in this book)

	$V_{a(r.m.s.)}$ (V)	C (μF)	R_{lim} min. (per anode) (Ω)
EZ80	2×250 2×300	50 50	125 215
EZ81	2×250 2×300	50 50	150 200
GZ34	2×400	8	125
UY85	200 250	60 60	90 100

CHAPTER 7
Three-watt Amplifier

The circuit described in this chapter has been developed to meet the demand for a simple amplifier of reasonably high quality. The amplifier, which is operated from a.c. mains, uses three Mullard valves: an EF86, an EL84 and an EZ80. The circuit given in Fig.1 includes three controls: volume (RV1), treble (RV2) and bass (RV11). A modified version of this circuit, which allows the amplifier to be used with a pre-amplifier or in stereophonic equipment, is shown in Fig.2. In this version, the three controls have been omitted, and the feedback network has been simplified.

The comparatively high sensitivity of the amplifier (100mV for 3W) permits the use of all types of crystal pick-up head and allows, if required, the use of equaliser networks between the head and amplifier. The output terminations of the circuit are suitable for almost all kinds of loudspeaker, but, although the circuit is designed to make the most effective use of the single output valve, the best possible results will only be achieved if a suitably housed high-quality speaker is used.

CIRCUIT DESCRIPTION

Because of the inherently high level of distortion with single-ended output stages, appreciable negative feedback around the output stage is necessary to produce an output of acceptable quality. At the same time, an overall sensitivity of 100mV is required if the amplifier is to be suitable for use with any type of crystal pick-up head. (The attenuation resulting from pick-up equalisation with positive RC networks must also be borne in mind.)

The basic sensitivity of the circuit without feedback should be about 10mV in order that the desirable level of feedback (about 20dB) can be provided. From considerations of stability, this feedback should be taken around the minimum number of stages.

Prototype of Three watt Amplifier

3W AMPLIFIER

The EF86 in the voltage-amplifying stage is used under conditions approaching those of starvation operation. With a high value of anode load resistance (R5 is 1MΩ) and reduced values of anode and screen-grid voltage, the gain of the stage is raised two or three times above that obtained under normal operating conditions. This increase is attributable mainly to the fact that, because the voltage at the anode of the EF86 is very low, direct coupling can be used between this anode and the control grid of the EL84 in the output stage. Thus the shunt loading on the anode circuit of the EF86 is least at low and medium frequencies.

The use of direct coupling between the stages necessitates a higher cathode voltage in the output stage than is required with RC coupling. The value of R13 is thus greater than is usual for the cathode resistance. The screen-grid voltage for the EF86 is taken from the cathode of the EL84. In this way, negative d.c. feedback (which is essential in a directly-coupled circuit to stabilise the operating conditions of both stages) is applied to the voltage amplifier.

Negative a.c. feedback is applied from the secondary winding of the output transformer to the cathode of the EF86. In Fig.1, this feedback loop incorporates the bass-boost control, the amount of feedback being

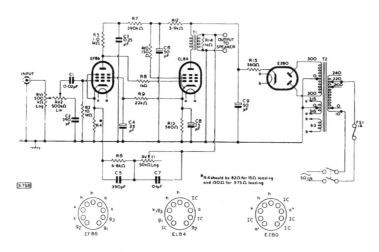

Fig. 1—*Complete circuit diagram of amplifier*

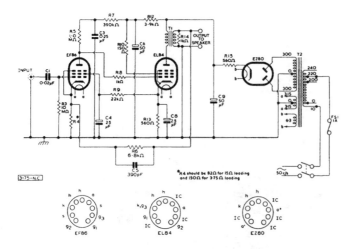

Fig. 2—*Circuit diagram of control-less amplifier*

changed continuously at low frequencies as the resistance of the control potentiometer RV11 is varied. In the simplified version of the circuit, the control RV11 is omitted, and the feedback loop consists of R6 and C5 only.

Provision for volume and treble control is made at the input of the amplifier. The potentiometers RV1 and RV2 constitute these controls respectively. In the control-less version of the amplifier, the output is taken directly to the capacitor C1 in the control-grid circuit of the EF86.

The power supply uses the EZ80 in combination with a mains transformer meeting the specification given below. The resistor R15 between the cathode of the EZ80 and the reservoir capacitor C9 is for voltage control. The anode of the EL84 is supplied from C9, and the screen-grid is supplied through another filter network R12, C6.

LIST OF COMPONENTS

Components indicated with an asterisk are not used in the control-less version of the amplifier.

Resistors

Circuit ref.	Value	Tolerance (±%)	Rating (W)
*RV1	500 kΩ	logarithmic potentiometer	
*RV2	500 kΩ	linear potentiometer	
R3	10 MΩ	20	¼
R4 for 15Ω speaker	82 Ω	10	¼
for 3·75Ω speaker	150 Ω	10	¼
[1]R5	1 MΩ	10	¼
R6	6·8 kΩ	10	¼
R7	390 kΩ	10	¼
R8	1 kΩ	20	¼
R9	22 kΩ	10	¼
R10	150 Ω	20	¼
*RV11	50 kΩ	logarithmic potentiometer	
R12	3·9 kΩ	10	½
[2]R13	560 Ω	5	3
R14	1 kΩ	20	¼
R15	560 Ω	20	2

[1] High stability, cracked carbon
[2] Wire wound

Output Transformer T1

Primary Impedance, 5kΩ

Commercial Components

Manufacturer	Type No.
Colne	03077
Elden	1264
Elstone	OT/3
Gardners	AS.7003
Gilson	W.O.767
Hinchley	1379
Parmeko	P2641
Partridge	P4073
Wynall	W1452

Valves

Mullard EF86, EL84, EZ80

Valveholders

B9A (noval) (two). McMurdo, BM9/U
B9A (noval) (one), nylon-loaded with screening skirt.
  McMurdo, XM9/AU, skirt 95

Capacitors

Circuit ref.	Value	Description	Rating (V)
C1	0·02 μF	paper	150
*C2	390 pF	silvered mica	
C3	0·25 μF	paper	350
C4	25 μF	electrolytic	50
C5	390 pF	silvered mica	
C6, C9	50+50 μF	double electrolytic	350
*C7	0·1 μF	paper	150
C8	25 μF	electrolytic	50

Tolerance of silvered mica capacitors is ±10%

Mains Transformer

Primary: 10–0–200–220–240V.
Secondaries: H.T. 300–0–300V, 60mA.
L.T. 3·15–0–3·15V, 1A (for EF86, EL84)
0–6·3V, 1A (for EZ80).

If only one 6·3V secondary winding is available, it should have a 2A rating to supply all three valves.

Commercial Components

Manufacturer	Type No.
Colne	03097
Elden	890A
Elstone	MT3M
Gardners	RS.3103
Gilson	W.O.839
Hinchley	1442
Parmeko	P2631
Partridge	H300/60
Wynall	W1547

Miscellaneous

Mains input plug, 3-pin. Bulgin, P340
Mains switch. N.S.F., 8280/K15
Mains voltage selector. Clix, CTSP/2
Fuse. Minifuse, Belling Lee, 1A
Lampholder (optional). Bulgin, D180/red
Pilot lamp (optional). Bulgin, 6·3V, 40mA
Input socket, coaxial. Aerialite, 149; Belling Lee, L.734/S
Output plug. 2-pin. Bulgin, P350
Tagboard. (10-way) Bulgin, C125; Denco

3W AMPLIFIER

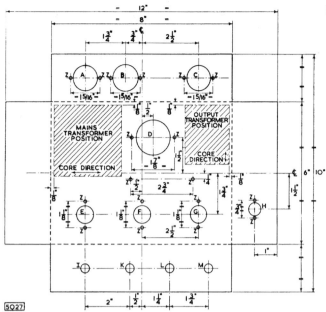

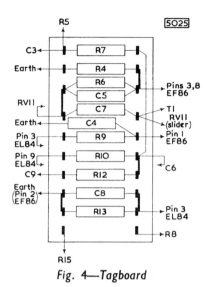

Fig. 3 (left)—Chassis details (The chassis should be bent up at 90° at all dotted lines)

Fig. 4—Tagboard

KEY TO HOLES IN CHASSIS			
Hole	Dimension	Use	Type No.
A	1 1/16 in. dia.	Mains input plug, 3-pin. Bulgin	P340
B	1 1/8 in. dia.	Voltage selector. Clix	CTSP/2
C	1 1/16 in. dia.	Output plug, 2-pin. Bulgin	P350
D	1 1/2 in. dia.	Electrolytic capacitor	—
E	3/4 in. dia.	B9A valveholder. McMurdo	BM9/U
F	3/4 in. dia.	B9A valveholder. McMurdo	BM9/U
G	3/4 in. dia.	B9A nylon-loaded valveholder with screening skirt. McMurdo	XM9/AU Skirt 95
H	1/2 in. dia.	Input socket, coaxial. Aerialite Belling Lee	149 L.734/5
I	3/8 in. dia.	Mains switch. N.S.F.	8280/K15
K	3/8 in. dia.	50kΩ logarithmic potentiometer	—
L	3/8 in. dia.	500kΩ linear potentiometer	—
M	3/8 in. dia.	500kΩ logarithmic potentiometer	—
Z	Drill No. 34	6 B.A. clearance hole	—

CONSTRUCTION AND ASSEMBLY

Chassis details of the amplifier are given in Fig. 3 For the control-less version, holes for the potentiometers will not be required. The chassis can be cut from one piece of 16 s.w.g. aluminium sheet, 12 in. long and 10 in. wide. A bottom cover plate to the amplifier is not necessary.

A suitable arrangement of the components in the amplifier with controls is shown in Fig. 5. The position of the components on the tagboard is shown in Fig. 4. For the control-less version, the potentiometers RV1, RV2 and RV11 and the capacitors C2 and C7 are omitted. The capacitor C1 will be connected directly to the input socket, and the feedback path will be completed by connecting the appropriate junction of R6 and C5 to the speaker terminal.

If the can of the double electrolytic capacitor is used as the negative side, then it should be isolated from the chassis. The earth connection to the chassis should be made at the input socket only.

The mains transformer should have an h.t. rating of 300-0-300V, 60mA, and it is preferable, though not essential, that a separate l.t. winding (6.3V) be used for the EZ80 rectifier. This is indicated in the circuit diagram, and also in the list of components.

PERFORMANCE

Frequency Response

With the treble and bass controls in Fig. 1 in their minimum effective positions, or with the control-less circuit of Fig. 2, the frequency response is essentially flat from 35c/s to 30kc/s (Fig. 6). With maximum application of the respective controls, a treble cut of 20dB is available at 10kc/s, and a bass boost of 15dB is available at 70c/s. The bass boost is obtained by reducing the main feedback at low frequencies by means of RV11 and C7 (Fig.1).

Distortion

The relationship between the total harmonic distortion and the output power is shown in Fig. 7. It will be seen that, for a typical amplifier, for outputs above about 3·5W, the distortion increases rapidly. This indicates the point beyond which over-loading of the amplifier occurs.

3W AMPLIFIER

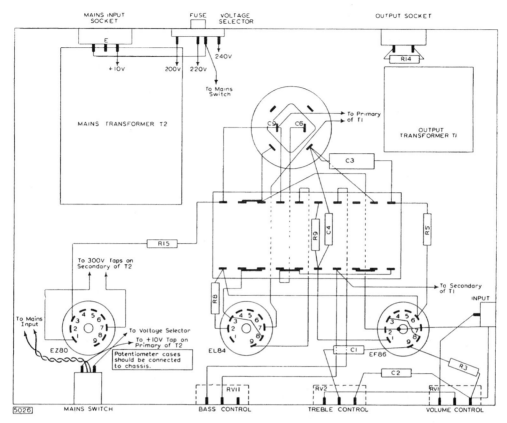

Fig. 5—Suggested layout of components

Underside of Prototype Amplifier

3W AMPLIFIER

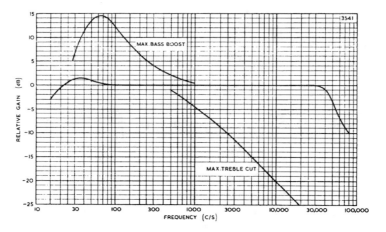

Fig. 6—Frequency response of amplifier showing relative gain without application of tone controls, and also showing relative gain with maximum application of tone controls

Output Impedance

The output impedance of the amplifier for a loudspeaker load of 15Ω is less than 1·5Ω. This gives an adequate damping factor of more than 10 (that is, >15/1·5).

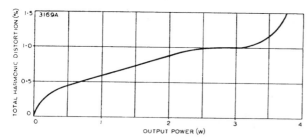

Fig. 7—Variation of total harmonic distortion with output power

D.C. CONDITIONS

The d.c. voltages at points in the equipment should be tested with reference to Table 1. The results shown in this table were obtained using an Avometer No. 8.

TABLE 1
D.C. Conditions

Point of Measurement		Voltages (V)	D.C. Range of Avometer* (V)
	C9	310	1000
	C6	290	1000
	C3	210	1000
EL84	Anode	290	1000
	Screen grid	290	1000
	Cathode	28	100
EF86	Anode	20	100
	Screen grid	28	100

*Resistance of Avometer
1000V-range, resistance = 20MΩ
100V-range, resistance = 2MΩ

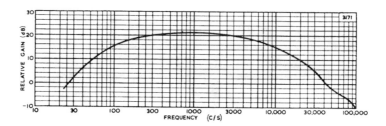

Fig. 8—Loop gain characteristics

CHAPTER 8

Seven-watt, D.C./A.C. Amplifier

The circuit described in this chapter is intended primarily for those who require high-quality audio reproduction but who are limited to, or are likely to encounter, d.c. mains supplies. The amplifier can be arranged to accommodate any speaker impedance without the need for rewiring if an output transformer having a tapped secondary winding is fitted. The absence of a mains transformer means that the unit is much smaller than other amplifiers of comparable output power.

The equipment operates in the range of mains voltage (a.c. or d.c.) of from 200 to 250V, and the maximum output power available is between 7 and 8W at low distortion. Negative feedback of 21dB is used to improve the performance of the complete amplifier, but even with this high value of feedback, the sensitivity of the circuit is high enough (200mV) to allow its use with most types of crystal pick-up ad or with f.m. radio tuner units.

Four Mullard B9A (noval) valves—types UF86, UCL82 (two) and UY85—are used in the amplifier. The heaters of these valves (rated at 100mA) are connected in a series chain. Because of this and the high a.c. voltages appearing at some of the heaters when the unit is operating on a.c. mains, the standard of performance in respect of hum pick-up is necessarily a little below that to be attained with equipment such as the 10 or 20W amplifiers. But the performance is still very good and, as satisfactory results have been obtained with a variety of makes of crystal pick-up head, the d.c./a.c. amplifier should satisfy most of the demands for equipment of this nature.

CIRCUIT DESCRIPTION

The circuit diagram for the d.c./a.c. amplifier is given in Fig. 1.

The input stage of the amplifier incorporates the Mullard high-gain, low-noise pentode, type UF86,

Prototype of D.C./A.C. Amplifier

D.C./A.C. AMPLIFIER

used as a voltage amplifier. In the second stage of the equipment, the triode sections of two triode-pentodes, type UCL82, form a phase-splitter. This provides the drive for the pentode sections of the UCL82s which operate as a self-biased, push-pull output stage with a nominal anode-to-cathode voltage of 200V.

Negative voltage feedback is used to improve the performance of the complete amplifier. It decreases distortion, improves the frequency response and reduces the output resistance.

The power supply of the amplifier uses the Mullard half-wave rectifier, type UY85. The rectifier output is RC smoothed, and the anodes of the push-pull pentodes are fed from the reservoir capacitor C20 through the centre-tapped primary winding of the output transformer.

Tone-control Circuit

The passive network comprising the tone-control circuit in this equipment is identical with that of the 10W amplifier described in Chapter 6. It will provide all the control necessary for normal usage.

Pick-up of hum voltages in the tone-control network is minimised by enclosing all the control components in a metal screening box. This box is isolated from the chassis by a sheet of paxolin and is held at the negative h.t. potential. The cases and spindles of the potentiometers are connected to this screening box. The chassis should not be connected to an earth point.

Details of the tone control are given in the circuit diagram in Fig. 1. The treble and bass controls are RV1 and RV3 respectively. RV5 is the volume control. The arrangement of the components within the metal screening box is given in Fig. 5.

Input Stage

The UF86 in the first stage of the amplifier is connected to the neutral end of the series heater chain. In this way, there is a minimum transfer of hum voltage from the heater to the control grid of the valve. To obtain a low level of noise in the amplifier, high-stability components should be used for the resistors R6, R7 and R9 in the anode, cathode and screen-grid circuits of the UF86.

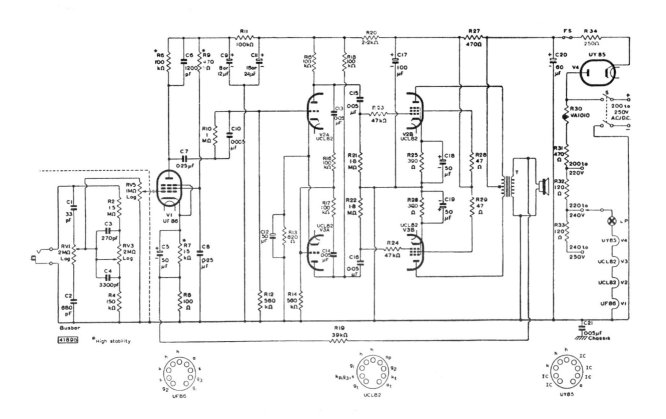

Fig. 1—Circuit diagram of d.c./a.c. amplifier

D.C./A.C. AMPLIFIER

Instability in the amplifier can result from the large amount of negative feedback introduced into the cathode circuit of the UF86. It can also occur because of the impedance that exists between the chassis and the busbar which serves as the negative return line of the amplifier. The following precautions are taken to prevent any possibility of instability.

The simplest method is to reduce the loop gain of the amplifier. To do this at high audio frequencies, the capacitor C6 is connected to bypass the anode load resistor R6. To reduce the loop gain at bass frequencies, the parallel combination of R10 and C10 is included in the coupling circuit between the first and second stages of the equipment. This method of

LIST OF COMPONENTS

Resistors

Circuit ref.	Value	Tolerance (±%)	Rating (W)
RV1	2 MΩ logarithmic potentiometer		
R2	1·5MΩ	10	¼
RV3	2 MΩ logarithmic potentiometer		
R4	150 kΩ	10	¼
RV5	1 MΩ logarithmic potentiometer		
*R6	100 kΩ	10	¼
*R7	1·5kΩ	10	¼
R8	100 Ω	10	¼
*R9	470 kΩ	10	¼
R10	1 MΩ	10	¼
R11	100 kΩ	10	¼
R12	560 kΩ	10	¼
R13	820 Ω	10	¼
R14	560 kΩ	10	¼
R15	100 kΩ	10	¼
R16	100 kΩ	5	¼
R17	100 kΩ	5	¼
R18	100 kΩ	10	¼
R19	39 kΩ	10	¼
R20	2·2kΩ	10	½
R21	1·8MΩ	10	¼
R22	1·8MΩ	10	¼
R23	47 kΩ	10	¼
R24	47 kΩ	10	¼
R25	390 Ω	10	2
R26	390 Ω	10	2
R27	470 Ω	10	1
R28	47 Ω	10	¼
R29	47 Ω	10	¼
R30	Varite thermistor, type VA1010		
†R31	470 Ω	10	6
R32	120 Ω	10	2
R33	120 Ω	10	2
R34	250 Ω	10	2

*High stability, cracked carbon
†Wire wound

Output Transformer

Primary Impedance, 6kΩ

Commercial Components

Manufacturer	Type No.
Colne	O3078
Elden	1263
Elstone	OT/6V
Gardners	AS.7008
Gilson	W.O.1059
Hinchley	6112
Parmeko	P2820
Partridge	P4081
Wynall	W1714

Capacitors

Circuit ref.	Value		Description	Rating (V)
C1	33	pF	silvered mica	—
C2	680	pF	silvered mica	—
C3	270	pF	silvered mica	—
C4	3300	pF	silvered mica	—
C5	50	μF	electrolytic	12
C6	1200	pF	silvered mica	—
C7	0·25	μF	paper	250
C8	0·25	μF	paper	250
‡C9	8 or 12	μF	electrolytic	250
C10	0·005	μF	paper	150
‡C11	16 or 24	μF	electrolytic	250
C12	50	μF	electrolytic	12
C13	0·05	μF	paper	250
C14	0·05	μF	paper	250
C15	0·05	μF	paper	250
C16	0·05	μF	paper	250
C18	50	μF	electrolytic	25
C19	50	μF	electrolytic	25
C20, C17	60+100	μF	electrolytic	250
C21	0·05	μF	paper	750

‡C9 and C11 comprise either an (8+8+8) or a (12+12+12) μF electrolytic capacitor.
Tolerance of all silvered mica capacitors is ±10%

Valves

Mullard UF86; UCL82 (two); UY85

Valveholders

B9A (noval) (three). McMurdo BM9/U

B9A (noval) (one), nylon-loaded with screening skirt. McMurdo XM9/AU, skirt 95

Miscellaneous

Mains input plug, 3-pin. Bulgin, P340
Mains switch, 2-pole. Bulgin, S300
Mains voltage selector. Clix CTSP/2
Fuseholder. Minifuse, Belling Lee, L.575
Fuse. Minifuse, Belling Lee, 150mA
Lampholder. Bulgin, D350
Pilot lamp. Bulgin, 6·3V, 110mA
Input jack. Igranic, P71
Output plug, 2-pin. Bulgin, P350
Tagboard (10-way (five). Bulgin, C125
Tagstrip. Bulgin, T94

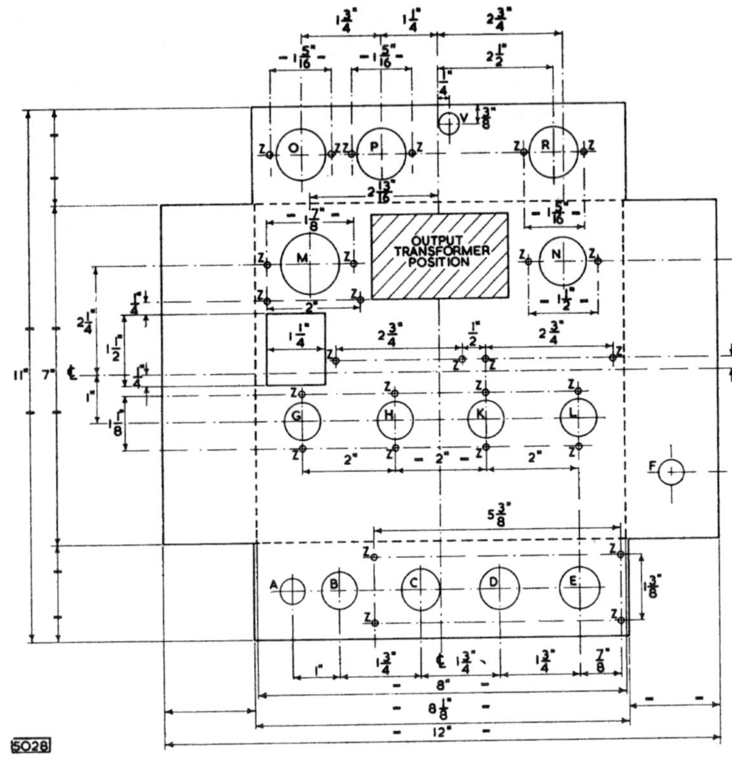

(*a*) *Main chassis*

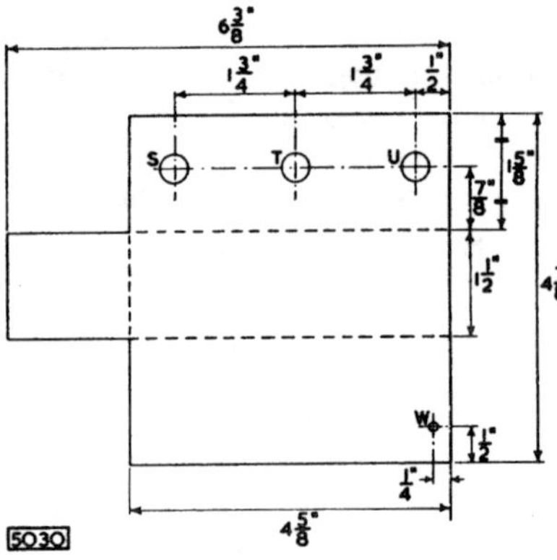

(*b*) *Shielding box*

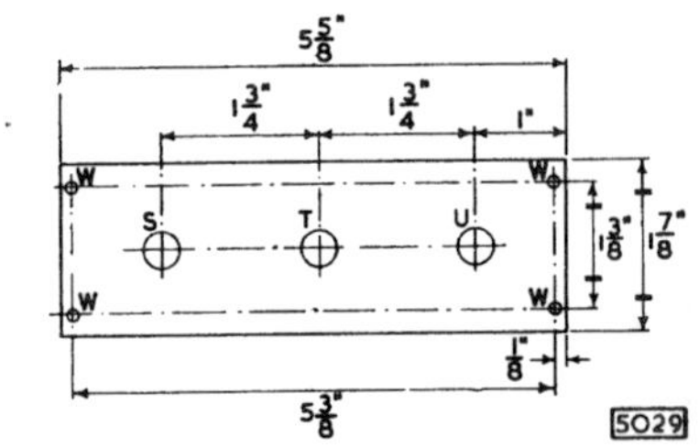

(*c*) *Insulator*

Fig. 2—Chassis details (the chassis should be bent up at 90° at all dotted lines)

KEY TO HOLES IN CHASSIS DRAWINGS

Hole	Dimension	Use	List No.
A	$\frac{1}{2}$ in. dia.	Mains switch, 2-pole. Bulgin.. ..	S300
B	$\frac{3}{4}$ in. dia.	Pilot lamp, Bulgin	D350
C	$\frac{7}{8}$ in. dia.	2MΩ potentiometer (logarithmic)..	—
D	$\frac{7}{8}$ in. dia.	2MΩ potentiometer (logarithmic)..	—
E	$\frac{7}{8}$ in. dia.	1MΩ potentiometer (logarithmic)..	—
F	$\frac{1}{2}$ in. dia.	Input jack, Igranic	P71
G	$\frac{3}{4}$ in. dia.	B9A valveholder, McMurdo ..	BM9/U
H	$\frac{3}{4}$ in. dia.	B9A valveholder, McMurdo ..	BM9/U
K	$\frac{3}{4}$ in. dia.	B9A valveholder, McMurdo ..	BM9/U
L	$\frac{3}{4}$ in. dia.	B9A nylon-loaded valveholder with screening skirt McMurdo ..	XM9/AU, Skirt 95
M	$1\frac{1}{4}$ in. dia.	Electrolytic capacitor	—
N	1 in. dia.	Electrolytic capacitor	—
O	$1\frac{1}{16}$ in. dia.	Mains input plug, 3-pin, Bulgin ..	P340
P	$1\frac{1}{16}$ in. dia.	Mains selector, Clix	CTSP/2
R	$1\frac{1}{16}$ in. dia.	Output plug, 2-pin, Bulgin ..	P350
S	$\frac{3}{8}$ in. dia	2MΩ potentiometer (logarithmic)..	—
T	$\frac{3}{8}$ in. dia	2MΩ potentiometer (logarithmic)..	—
U	$\frac{3}{8}$ in. dia	1MΩ potentiometer (logarithmic)..	—
V	$\frac{7}{16}$ in dia	Fuseholder, Minifuse Belling Lee ..	L.575
W	Drill No. 31	—	—
Z	Drill No. 34	6BA clearance holes	—

coupling results in resistive attenuation at the low audio frequencies, so that the phase shift is limited. In addition, the busbar is connected to the chassis by way of the capacitor C21.

Phase-splitting Stage

The triode sections of the UCL82s form an anode-follower phase-splitting stage. Such an arrangement has been chosen because of the large amount of negative feedback that occurs between the anode and grid of V3A. This can be used to counteract the large hum voltages transferred, with an a.c. mains supply, to the grid of V3A as a result of running the heater of V3 some 62V above the neutral voltage level.

The balance between the output voltages from the triodes of this stage is controlled by the values chosen for the resistors R14, R16 and R17. If resistors which deviate from the nominal values by the extreme of the 5% tolerance range are used for R16 and R17, the resulting lack of balance should not be more than 4%.

D.C./A.C. AMPLIFIER

Output Stage

The pentode sections of the UCL82s are used in a push-pull output stage. Normal pentode loading is used so that maximum output power is obtained with any h.t. voltage. The valves operate under the following Class AB conditions:

Anode-to-cathode voltage	200V
Screen-to-cathode voltage	200V
Cathode circuit resistance (each pentode)	390Ω
Anode-to-anode load	6kΩ
Quiescent anode current	2×35mA

Rectifier Stage

The resistor R34 is included in the cathode circuit of the UY85 to limit surge currents through the valve. The fuse FS, rated at 150mA, is also included in the h.t. circuit.

Heater Circuit

The heaters of the four valves used in the amplifier are connected in series. The UF86 in the input stage (which has the highest gain) is connected to the neutral side of the mains supply.

The nominal drop in voltage across the four heaters in series is 150V. The difference between this and a mains voltage in the range 200 to 250V is taken up by the following components connected in the heater circuit: the Varite thermistor R30 (type VA1010) used for surge suppression; the resistors R31 (470Ω), R32 (120Ω) and R33 (120Ω) used to provide tapping points for the ranges of mains voltage; the pilot lamp LP used as an indicator. The components R30, R31 and R32 are mounted under a ventilation hole cut in the chassis. This allows dissipation of the heat developed in these resistors.

Negative Feedback

The amount of feedback taken from the secondary winding of the transformer to the cathode circuit of the UF86, is 21dB. The output resistance with this feedback is 0·9Ω measured at the 15Ω output terminals. This gives an adequate damping factor of about 17.

CONSTRUCTION AND ASSEMBLY

Chassis details for the d.c./a.c. amplifier are given in Fig. 2. The two metal pieces of the chassis should be cut from 16 s.w.g. aluminium sheet and the insulator from paxolin sheet, $\frac{1}{16}$ in. thick. The dimensions (in inches) of these pieces are:

Main chassis	12×11
Screening box	$6\frac{3}{8} \times 4\frac{7}{8}$
Insulator	$5\frac{5}{8} \times 1\frac{7}{8}$

The arrangements of the components on the tagboards is given in Figs. 3 and 4 and a suggested layout of the components in the chassis is shown in Fig. 5.

In all d.c./a.c. equipment, certain precautions must be taken to prevent constructors and users receiving electrical shocks. This possibility of shocks arises because the amplifier can inadvertently be connected wrongly to the mains supply. If the polarity of the mains connection is incorrect, several components will become 'live'.

In this amplifier, the chassis is insulated electrically from the internal components and wiring except for a single connection, by way of C21, between the chassis and the busbar return line of the amplifier. (This connection is required for reasons of stability.)

The amplifier itself should be contained in a well-ventilated insulating box. If this is not possible then at least the ventilation hole and the electrolytic capacitors on the top of the chassis must be protected to prevent people touching them. The potentiometers should be fitted with knobs which prevent any contact with the bare spindles and which have no protruding grub screws.

To prevent hum pick-up, there are no isolating capacitors in the input circuit. For this reason great care must be taken when connecting auxiliary equipment to the amplifier. It is essential that no exposed metal parts are connected to either side of the input circuit of the amplifier. Earth connections in associated equipment should be dispensed with, as there is a risk that one side of the mains may become connected to them. They are also likely to cause hum trouble.

It should also be borne in mind that the speaker leads are connected to the busbar, and because of an incorrect mains connection this may become 'live'. Thus the metal parts of the loudspeaker should be inaccessible.

PERFORMANCE

The performance data given in this chapter have been obtained from measurements made on a prototype amplifier using a typical output transformer. Except for the relationship between output power and mains voltage, all the data have been obtained with an a.c. mains voltage of 240V. The measurements have been made with a sine-wave input signal.

Output Power

The variation of obtainable output power with mains voltage is shown in Fig. 6. It can be seen that the rated output of 7W at an input frequency of 1kc/s is obtainable with an a.c. mains voltage of 240V.

D.C./A.C. AMPLIFIER

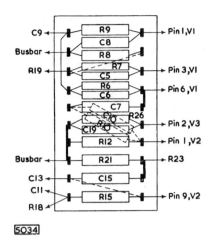

Fig. 3—Tagboard No. 1

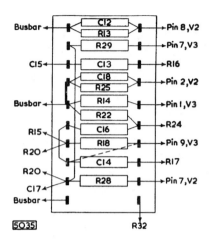

Fig. 4—Tagboard No. 2

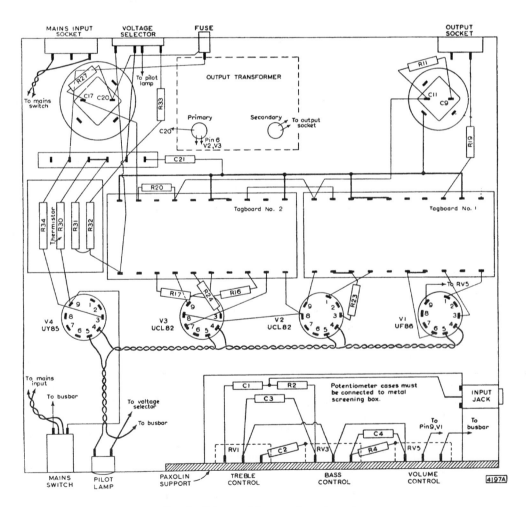

Fig. 5—Suggested layout of components

D.C./A.C. AMPLIFIER

The maximum output for a mains voltage of 250V is 8W. When a d.c. mains supply is used, the output power will be lower because less h.t. voltage will be available.

Distortion

The total harmonic distortion, measured at 400c/s, is about 0·5% for an output of 7W. For an output of 3W, the distortion is 0·2%.

Intermodulation distortion was measured at 40c/s and 10kc/s, the amplitude of the low-frequency signal being four times that of the high-frequency signal. At an equivalent output power of 7W, the intermodulation distortion is 2%. At an equivalent power of 3W, the distortion is 1%. The stipulated equivalent output power is the power that would be obtained from a single sine-wave input of amplitude equal to the peak value of the combined low- and high-frequency inputs. The distortion is expressed in r.m.s. terms as the percentage of the intermodulation products relative to the higher frequency.

The magnitude of the beat-note distortion was found to be negligible in the prototype.

Frequency Response and Stability

The frequency-response, loop-gain and phase-shift characteristics for the prototype amplifier are given in Fig. 7.

At the full rated output of 7W, the response curve is flat to within ±1dB (relative to the output level at 1kc/s) for frequencies ranging from 15c/s to 14kc/s.

TABLE 1
D.C. Conditions

Point of Measurement		Voltages (V)	D.C. Range of Avometer* (V)
	C20	217	1000
	C17	210	1000
	C11	203	1000
	C9	128	1000
UCL82 V3B, V2B	Anodes	215	1000
	Screen grids	209	1000
	Cathodes	15·8	100
V3A, V2A	Anodes	116	1000
	Cathodes	1·5	10
UF86 V1	Anode	61	100
	Screen grid	60	100
	Cathode	1·3	10

*Resistance of Avometer:
1000V-range, resistance = 20MΩ
100V-range, resistance = 2MΩ
10V-range, resistance = 200kΩ

Underside View of Prototype Amplifier

D.C./A.C. AMPLIFIER

The phase shift in the feedback loop is only 117° at a frequency of 22kc/s when the loop gain is unity. Consequently, the amplifier has a stability margin at high frequencies of 63°. The phase-shift characteristic below 100c/s is not given in Fig. 7. Nevertheless, adequate stability at low frequencies has been achieved in the amplifier.

Tone Controls

The treble tone-control potentiometer RV1 gives continuously variable control from +10dB to −10dB at 10kc/s, and the bass potentiometer RV3 gives such control from +8dB to −5dB at 100c/s.

Sensitivity

The input voltage required for the amplifier (including the tone control network) to give the rated output of 7W is 200mV. This is enough to allow its use with most types of crystal pick-up head or f.m. tuner unit.

Hum and Noise

In several copies of the prototype used with a.c. mains supplies, the hum and noise level, with the volume control set to zero, was always better than 60dB below 7W. This is not as good as the level normally achieved with present-day high-quality audio equipment. But the level is still very satisfactory, and can be considered good if the limitations set by the series heater chain and the associated high alternating potentials with a.c. mains operation are borne in mind.

D.C. CONDITIONS

The d.c. voltages at points in the equipment should be tested with reference to Table 1. The results shown in this table were obtained using an Avometer No. 8.

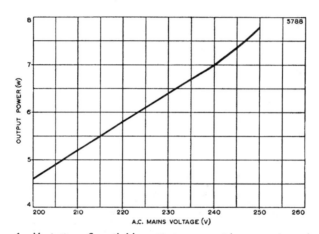

Fig. 6—Variation of available output power with a.c. mains voltage

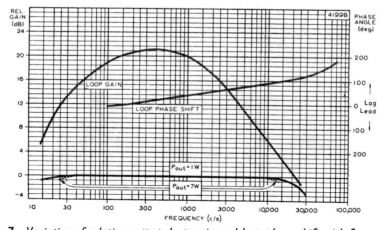

Fig. 7—Variation of relative output, loop gain and loop phase shift with frequency

CHAPTER 9
Two- and Three-valve Pre-amplifiers

The circuits described in this chapter have been designed principally for use with amplifiers built to the 20, 10, or 3W circuits discussed in Chapters 5, 6 and 7. They can, however, be used with any power amplifier which does not require an input signal greater than 250mV for full output. Input facilities are provided for magnetic and crystal pick-up heads, tape-recorder playback heads, radio tuner units and (in the 2-valve circuit) a microphone. An auxiliary input position is provided for ancillary equipment such as the tape pre-amplifier circuit described in Chapter 12.

An additional output position, which is independent of the tone controls, is also provided with both circuits, enabling programmes to be taken to a tape amplifier while they are being fed from the normal output position to a power amplifier. Both the auxiliary input and the additional output positions have jack sockets which are situated at the front of the chassis.

All the input sockets in each circuit are connected to one switch which selects one input at a time. In both circuits this switch also short-circuits the unused sockets to earth, an arrangement which considerably reduces the amount of 'break-through' between channels. The positions of the switch, from left to right, are: Auxiliary, Radio, Tape, Microphone (in the 2-valve circuit only) Microgroove and 78 r.p.m.

The equalisation for disc recordings conforms to the present R.I.A.A. characteristics which have been adopted by most of the major recording companies. The tape-playback characteristic is intended for use with high-impedance heads when replaying pre-recorded tapes at a speed of $7\frac{1}{2}$ inches per second.

The tone controls used in both circuits cover a wide range of frequencies and provide boost and cut for high and low frequencies. Switched high- and low-pass filter networks are included in the 3-valve circuit so that unwanted signals such as rumble and record scratch can be eliminated.

Prototype of 2-valve Pre-amplifier

Prototype of 3-valve Pre-amplifier

2- AND 3-VALVE PRE-AMPLIFIERS

CIRCUIT DESCRIPTIONS
Two Mullard high-gain pentodes, type EF86, are used in the 2-valve circuit of Fig. 2. In the 3-valve circuit shown in Fig. 3, two EF86s and a Mullard double triode, type ECC83, are used.

First Stage of 2- and 3-Valve Circuits
All the equalisation takes place in the first stage of both circuits, and is achieved by means of frequency-selective feedback between the anode and grid of the first EF86. This arrangement has been chosen because the grid-circuit impedance of the first stage should be low. A low impedance at this grid lessens hum pick-up and reduces the effect of plugging-in external low-impedance circuits. Furthermore, the arrangement also results in low gain in the first stage. Hence, Miller effect between the anode and grid of the first EF86, which can be troublesome when high values of resistance are used in series with the grid, is reduced.

Series resistors are used in the input circuits so that the sensitivity and impedance of any channel can be adjusted accurately. The component values given in Figs. 2 and 3 are intended for sources encountered most frequently, but the sensitivity and impedance[1] of each channel can be altered by changing the value of the appropriate series resistor.

---

[1] The impedance of the input channels includes the grid impedance of the EF86 modified by the feedback components as well as the impedance of the input network.

Second Stage of 2-valve Circuit
No feedback is used in the second stage of the 2-valve pre-amplifier. The full output from the anode of the second EF86 is taken to the passive tone-control network and an auxiliary output is taken from the junction of R20 and R21 to the programme-recording jack socket.

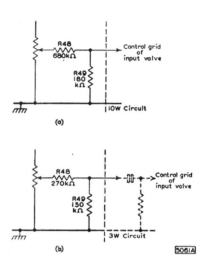

Fig. 1—Output-attenuating networks for use with **both** pre-amplifiers

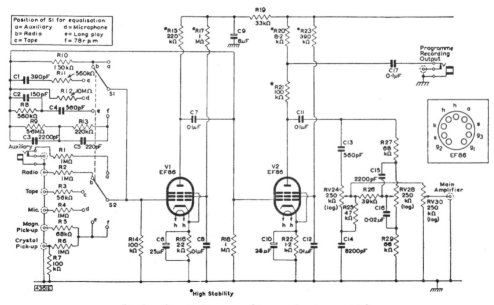

Fig. 2—Circuit diagram of **two-valve** pre-amplifier

2- AND 3-VALVE PRE-AMPLIFIERS

Second Stage of 3-valve Circuit

The second stage of the 3-valve pre-amplifier circuit of Fig. 3 has a linear frequency-response characteristic and its gain is reduced by a small amount of negative feedback applied at the control grid by way of the resistor R23. The output of this stage is taken from the anode to the tone-control network, and also to the programme-recording output socket. Because of the negative feedback, the tone controls have very little effect on the frequency response at this anode.

Third Stage of 3-valve Circuit

The output stage of the 3-valve pre-amplifier is made up of the filter circuits. The low-pass filter is situated between the two sections of the ECC83, and is arranged on the switch SB. It incorporates a Mullard wound pot-core inductor, type WF1428, in a π-type network. The high-pass filter is arranged on the switch SC and consists of two RC networks in a feedback loop around the second triode section of the valve.

Output Attenuating Networks

The sensitivity and impedance of each channel have been chosen for sources which are likely to be encountered most frequently. The level of output signal from the pre-amplifier will have to be adjusted if it is to be used with power amplifiers having different sensitivities from the 20W circuit. The signal should be attenuated by using a simple potential divider as shown in Fig. 1. The values of resistance in Fig. 1 (*a*) are for an attenuator suitable for the control-less version of the 10W amplifier (Chapter 6) and those in Fig. 1 (*b*) are suitable for the control-less 3W amplifier (Chapter 7).

LIST OF COMPONENTS FOR TWO-VALVE PRE-AMPLIFIER

Resistors

Circuit ref.	Value	Tolerance (±%)	Rating (W)
R1	1 MΩ	10	¼
R2	1 MΩ	10	¼
R3	56 kΩ	10	¼
R4	1 MΩ	10	¼
R5	68 kΩ	10	¼
R6	1 MΩ	10	¼
R7	100 kΩ	10	¼
R8	560 kΩ	10	¼
R9	5·6 MΩ	10	¼
R10	150 kΩ	10	¼
R11	560 kΩ	10	¼
R12	10 MΩ	10	¼
R13	220 kΩ	10	¼
R14	100 kΩ	10	¼
[1]R15	220 kΩ	10	½
R16	2·2 kΩ	10	¼
[1]R17	1 MΩ	10	½
R18	1 MΩ	10	¼
R19	33 kΩ	10	½
[1]R20	8·2 kΩ	10	½
[1]R21	100 kΩ	10	½
R22	1·2 kΩ	10	¼
[1]R23	390 kΩ	10	½
RV24	250 kΩ	logarithmic potentiometer	
R25	47 kΩ	10	¼
R26	39 kΩ	10	¼
R27	68 kΩ	10	¼
RV28	250 kΩ	logarithmic potentiometer	
R29	6·8 kΩ	10	¼
RV30	250 kΩ	logarithmic potentiometer	
R48 (Fig. 1a)	680 kΩ	10	¼
R48 (Fig. 1b)	270 kΩ	10	¼
R49 (Fig. 1a)	180 kΩ	10	¼
R49 (Fig. 1b)	150 kΩ	10	¼

1. High stability, cracked carbon

Valves

Mullard EF86 (two)

Capacitors

Circuit ref.	Value	Description	Rating (V)
C1	390 pF	silvered mica	
C2	150 pF	silvered mica	
C3	2200 pF	silvered mica	
C4	560 pF	silvered mica	
C5	220 pF	silvered mica	
C6	25 µF	electrolytic	12
C7	0·1 µF	paper	350
C8	0·1 µF	paper	350
C9	8 µF	electrolytic	350
C10	25 µF	electrolytic	12
C11	0·1 µF	paper	350
C12	0·1 µF	paper	350
C13	560 pF	silvered mica	
C14	8200 pF	silvered mica	
C15	2200 pF	silvered mica	
C16	0·02 µF	paper	
C17	0·1 µF	paper	350

Tolerance of silvered mica capacitors is 10%

Valveholders

B9A (noval) nylon-loaded, with screening skirt.
McMurdo XM9/AU, skirt 95 (two).

Miscellaneous

Supply input socket, 6-pin. Bulgin, P194
Input socket (six), coaxial. Belling Lee, L.604/S
Input jack. Igranic, P71
Output socket, coaxial. Belling Lee, L.604/S
Output jack. Igranic, P71
Selector switch, 6-way rotary.
  Shirley Laboratories, SBL M/S/190
  Specialist Switches, SS/592
(Note: Details of proprietary switches may not be identical with those shown in the diagram.)
Tagboard, 10-way (two). Bulgin, C125; Denco

2- AND 3-VALVE PRE-AMPLIFIERS

LIST OF COMPONENTS FOR THREE-VALVE PRE-AMPLIFIER

Resistors

Circuit ref.	Value	Tolerance (±%)	Rating (W)
R1	1 MΩ	10	¼
R2	1 MΩ	10	¼
R3	56 kΩ	10	¼
R4	100 kΩ	10	¼
R5	2·2 MΩ	10	¼
R6	100 kΩ	10	¼
R7	8·2 MΩ	10	¼
R8	2·2 MΩ	10	¼
R9	100 kΩ	10	¼
R10	390 kΩ	10	¼
R11	270 kΩ	10	¼
R12	180 kΩ	10	¼
R13	100 kΩ	10	¼
[1]R14	270 kΩ	10	½
R15	3·9 kΩ	10	¼
[1]R16	1·5 MΩ	10	½
R17	33 kΩ	10	¼
R18	220 kΩ	10	¼
R19	1 MΩ	10	¼
[1]R20	100 kΩ	10	½
R21	1·2 kΩ	10	¼
[1]R22	470 kΩ	10	½
R23	3·9 MΩ	10	¼
RV24	250 kΩ logarithmic potentiometer		
R25	47 kΩ	10	¼
R26	39 kΩ	10	¼
R27	68 kΩ	10	¼
RV28	250 kΩ logarithmic potentiometer		
R29	6·8 kΩ	10	¼
R30	270 kΩ	10	¼
R31	33 kΩ	10	¼
R32	22 kΩ	10	¼
R33	1·2 kΩ	10	¼
R34	10 kΩ	10	¼
R35	22 kΩ	10	¼
R36	12 kΩ	10	¼
R37	270 kΩ	10	¼
R38	33 kΩ	10	¼
R39	47 kΩ	10	¼
R40	56 kΩ	10	¼
R41	56 kΩ	10	¼
R42	1·5 MΩ	10	¼
R43	220 kΩ	10	¼
R44	47 kΩ	10	½
R45	1·5 kΩ	10	½
R46	220 kΩ	10	¼
RV47	50 kΩ logarithmic potentiometer		
R48 (Fig. 1a)	680 kΩ	10	¼
R48 (Fig. 1b)	270 kΩ	10	¼
R49 (Fig. 1a)	180 kΩ	10	¼
R49 (Fig. 1b)	150 kΩ	10	¼

1. High stability, cracked carbon

Inductor, L1

Wound Ferroxcube pot core, Mullard type WF1428

Valves

Mullard EF86 (two), ECC83

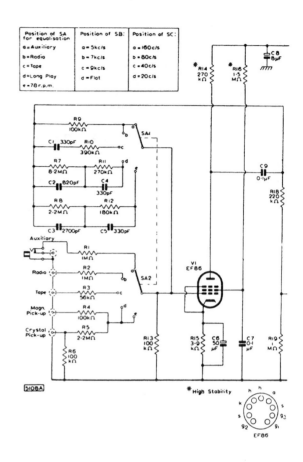

Fig. 3—Circuit diagram

Valveholders
B9A (noval) nylon-loaded, with screening skirt
 McMurdo, XM9/AU, Skirt 95 (three)

Miscellaneous
Supply input socket, 6-way. Bulgin, P194
Lampholder. Bulgin, D/675/1/Red
Pilot lamp. 6·3V, 150mA
Input socket (four), coaxial. Belling Lee, S.604/S
Input jack. Igranic, P71
Output socket, coaxial. Belling Lee, S.604/S
Output jack. Igranic, P71
Selector switch SA, 5-way rotary
 Specialist Switches, SS/593/A
Low-pass filter switch SB, 4-way rotary
 Specialist Switches, SS/593/B
High-pass filter switch SC, 4-way rotary
 Specialist Switches, SS/593/C
Set of 3 switches. Shirley Laboratories, SBL F/5/62
(Note: Details of proprietary switches may not be identical with those given in the diagrams.)
Tagboard, 5-way (one). Bulgin, C120
Tagboard, 10-way (three). Bulgin, C125

2- AND 3-VALVE PRE-AMPLIFIERS

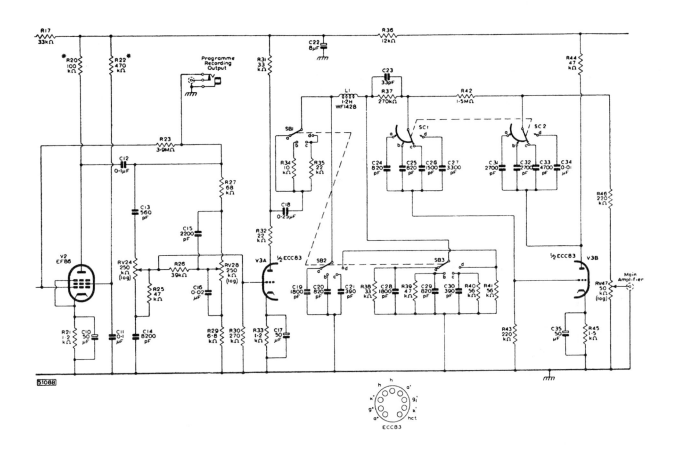

of **three-valve** *pre-amplifier*

Capacitors

Circuit ref.	Value		Description	Rating (V)	Circuit ref.	Value		Description	Rating (V)
					C18	0.25	μF	paper	350
C1	330	pF	silvered mica		C19	1800	pF	silvered mica	
C2	820	pF	silvered mica		C20	820	pF	silvered mica	
C3	2700	pF	silvered mica		C21	390	pF	silvered mica	
C4	330	pF	silvered mica		C22	8	μF	electrolytic	350
C5	330	pF	silvered mica		C23	33	pF	silvered mica	
C6	50	μF	electrolytic	12	C24	820	pF	silvered mica	
C7	0.1	μF	paper	350	C25	820	pF	silvered mica	
C8	8	μF	electrolytic	350	C26	1500	pF	silvered mica	
C9	0.1	μF	paper	350	C27	3300	pF	silvered mica	
C10	50	μF	electrolytic	12	C28	1800	pF	silvered mica	
C11	0.1	μF	paper	350	C29	820	pF	silvered mica	
C12	0.1	μF	paper	350	C30	390	pF	silvered mica	
C13	560	pF	silvered mica		C31	2700	pF	silvered mica	
C14	8200	pF	silvered mica		C32	2700	pF	silvered mica	
C15	2200	pF	silvered mica		C33	4700	pF	silvered mica	
C16	0.02	μF	paper		C34	0.01	μF	silvered mica	
C17	50	μF	electrolytic	12	C35	50	μF	electrolytic	12

Tolerance of silvered mica capacitors is 10%

2- AND 3-VALVE PRE-AMPLIFIERS

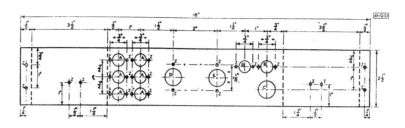

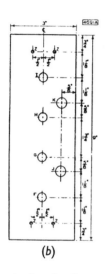
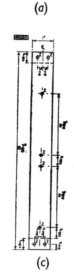
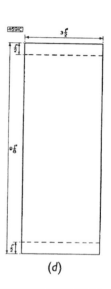

Fig. 4—Chassis details of **two-valve** pre-amplifier. (The chassis sections should be bent up at 90° at all dotted lines)

(a) Rear panel
(b) Front panel
(c) Mounting strip
(d) Cover panel (two)

KEY TO HOLES IN CHASSIS (Two-valve pre-amplifier)			
Hole	Dimension	Use	Type
A	½ in. dia.	Input socket, coaxial. Belling Lee	L.604/S
B	½ in. dia.	Output socket, coaxial. Belling Lee	L.604/S
C	¾ in. dia.	Supply input socket, 6-way. Bulgin	P194
D	¾ in dia	B9A nylon-loaded valveholder with screening skirt. McMurdo	XM9/AU Skirt 95
E	¾ in dia	B9A nylon-loaded valveholder with screening skirt. McMurdo	XM9/AU Skirt 95
F	⅜ in dia	Selector switch, 6-way	—
G	⅜ in. dia.	250kΩ potentiometer (logarithmic)	—
H	⅜ in. dia.	250kΩ potentiometer (logarithmic)	—
I	⅜ in. dia.	250kΩ potentiometer (logarithmic)	—
Z	Drill No. 34	6B.A. clearance hole	—

TABLE 1
H.T. Smoothing Components for 2-valve Pre-amplifier

Power Amplifier	Smoothing Resistor (kΩ±10%; ½W)	Decoupling Capacitor (μF)
20-watt	56	16
10-watt	22	16
3-watt	22	16

TABLE 2
H.T. Smoothing Components for 3-valve Pre-amplifier

Power Amplifier	Smoothing Resistor (kΩ±10%; ½W)	Decoupling Capacitor (μF)
20-watt	33	16
10-watt	12	16
3-watt	12	16

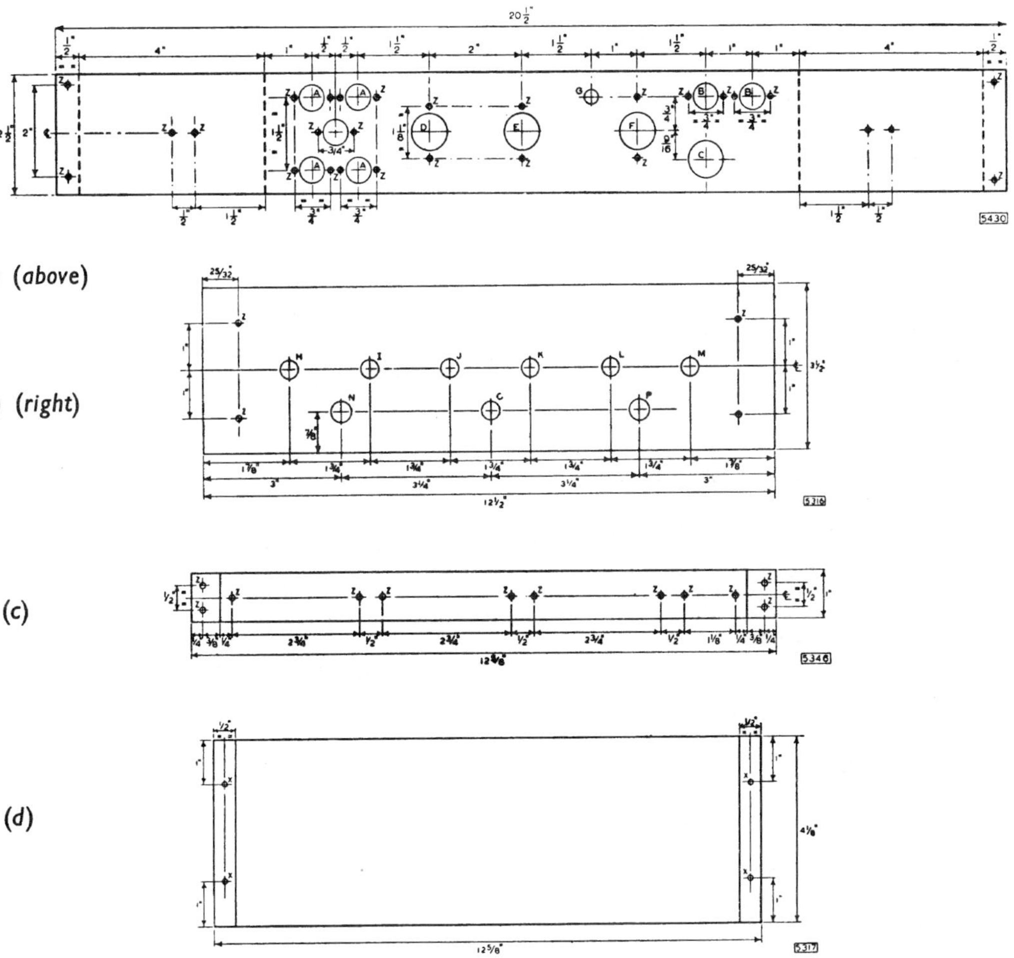

(a) (above)

(b) (right)

(c)

(d)

Fig. 5—Chassis details of ***three-valve*** *pre-amplifier. (The chassis sections should be bent up at 90° at all dotted lines)*

(a) Rear panel
(b) Front panel
(c) Mounting strip
(d) Cover panel (two)

KEY TO HOLES IN CHASSIS
(Three-valve pre-amplifier)

Hole	Dimension	Use	Type
A	½ in. dia.	Input socket, coaxial. Belling Lee ..	L.604/S
B	½ in. dia.	Output socket, coaxial. Belling Lee..	L.604/S
C	¾ in. dia.	Supply input socket, 6-way. Bulgin ..	P194
D	¾ in. dia.	B9A nylon-loaded valveholder with screening skirt. McMurdo.. ..	XM9/AU Skirt 95
E	¾ in. dia.	B9A nylon-loaded valveholder with screening skirt. McMurdo.. ..	XM9/AU Skirt 95
F	¾ in. dia.	B9A nylon-loaded valveholder with screening skirt. McMurdo.. ..	XM9/AU Skirt 95
G	5/16 in. dia.	Ferroxcube pot core	LA42
H	⅜ in. dia.	Selector switch, 5-way	—
I	⅜ in. dia.	250kΩ potentiometer (logarithmic)..	—
J	⅜ in. dia.	250kΩ potentiometer (logarithmic)..	—
K	⅜ in. dia.	Filter switch, 4-way	—
L	⅜ in. dia.	50kΩ potentiometer (logarithmic) ..	—
M	⅜ in dia.	Filter switch, 4-way	—
N	7/16 in. dia.	Input jack. Igranic	P71
O	⅜ in. dia.	Lampholder	—
P	7/16 in. dia.	Output jack. Igranic	P71
Y	Drill No. 49	Self-tapping screw	½in.×4
Z	Drill No. 34	6 B.A. clearance hole	—

H.T. Supplies

The h.t. currents taken by the 2-valve and 3-valve pre-amplifiers are respectively 3mA at 230V and 6mA at 250V. The values of the smoothing components for these supplies depend on the combination of pre-amplifier and power amplifier. Appropriate values are given in Tables 1 and 2. The points from which these supplies are taken are shown in the complete circuit diagrams of Chapters 5 and 6 for the 20W and 10W amplifiers. The point of connection for the smoothing resistors in the 3W amplifier circuit of Chapter 7 should be the reservoir capacitor C9.

The l.t. current taken by the 2-valve pre-amplifier is 0·4A at 6·3V and that required by the 3-valve circuit is 0·7A at 6·3V.

2- AND 3-VALVE PRE-AMPLIFIERS

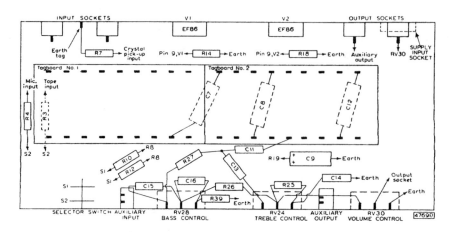

*Fig. 6—Suggested layout of components for **two-valve** circuit*

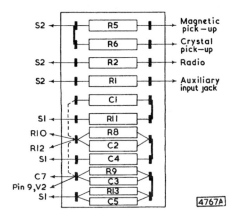

*Fig. 7—Tagboard No. 1, **two-valve** circuit*

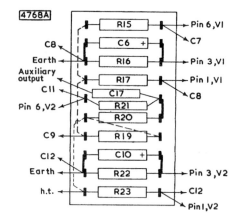

*Fig. 8—Tagboard No. 2, **two-valve** circuit*

CONSTRUCTIONAL DETAILS

The chassis design and layout of both pre-amplifiers are intended specifically for the home constructor. Conventional box-type chassis are not used. Instead the chassis are made up on the unit system, the separate parts being joined together during the assembly of the equipment. Both chassis are made up of five separate pieces of 16 s.w.g. aluminium sheet, the dimensions (in inches) of which are as follows:

	2-valve pre-amplifier	3-valve pre-amplifier
(a) Front panel	9×3	$12\frac{1}{2} \times 3\frac{1}{2}$
(b) Rear panel	$16 \times 2\frac{1}{2}$	$20\frac{1}{2} \times 2\frac{1}{2}$
(c) Mounting strip	$8\frac{7}{8} \times 1$	$12\frac{5}{8} \times 1$
(d) Cover panel (two)	$9\frac{1}{8} \times 3\frac{1}{2}$	$12\frac{5}{8} \times 4\frac{1}{8}$

The pieces of the chassis should be marked and cut as indicated in Fig. 4 (2-valve circuit) and Fig. 5 (3-valve circuit).

For ease of assembly, components should be mounted on the tagboards, and these should then be bolted to the mounting strip before the strip is attached to the rear plate. Diagrams showing the positions of the components on the tagboards are given in Figs. 7 and 8 for the 2-valve pre-amplifier, and in Figs. 10, 11, 12 and 13 for the 3-valve circuit.

When the mounting strip and tagboards have been attached to the rear plate, connections should be made between the valveholders and components. The valveholders should be mounted so that the valves will be on the outside of the completed chassis and the valves should be fitted with screening cans.

The potentiometers and switches should be mounted on the front panel of the chassis and the other components which make up the tone-control network should be connected to them. The front panel should

2- AND 3-VALVE PRE-AMPLIFIERS

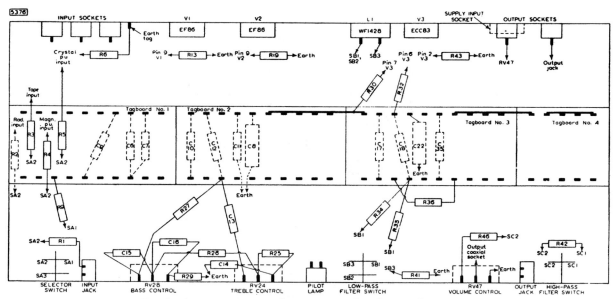

*Fig. 9—Suggested layout of components for **three-valve** circuit*

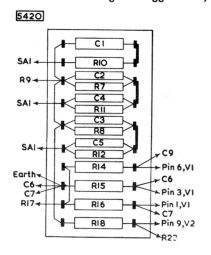

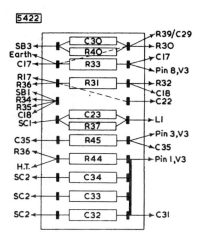

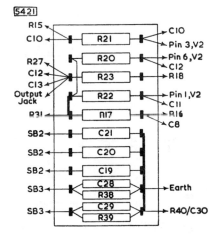

*Fig. 10—(left upper) Tagboard No 1, **three-valve** circuit*

*Fig. 11—(left lower) Tagboard No. 2, **three-valve** circuit*

*Fig. 12—(above) Tagboard No. 3, **three-valve** circuit*

*Fig. 13—(below) Tagboard No. 4, **three-valve** circuit*

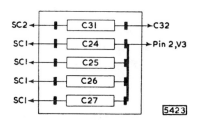

2- AND 3-VALVE PRE-AMPLIFIERS

*Top View of Prototype **2-valve** Pre-amplifier*

*Underside View of Prototype **2-valve** Pre-amplifier*

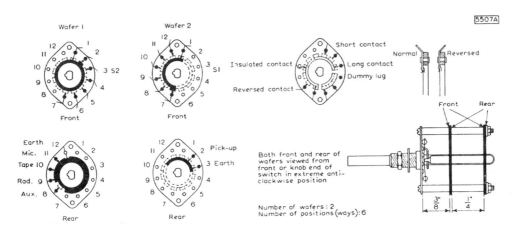

*Fig. 14—Selector-switch details for **two-valve** pre-amplifier*

2- AND 3-VALVE PRE-AMPLIFIERS

Top View of Prototype **3-valve** *Pre-amplifier*

Underside View of Prototype **3-valve** *Pre-amplifier*

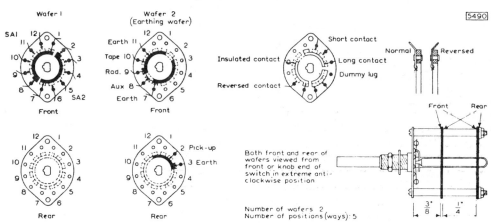

Fig. 15—*Selector-switch details for* **three-valve** *pre-amplifier*

2- AND 3-VALVE PRE-AMPLIFIERS

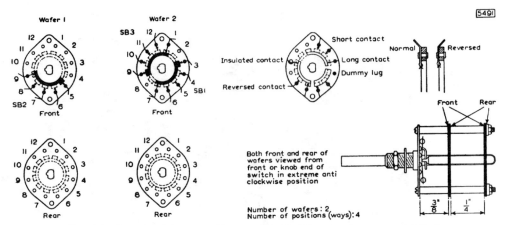

Fig. 16—Details of low-pass filter switch SB (**three-valve** pre-amplifier)

then be bolted to the back panel, and the remaining components should be connected in position as indicated in the general layout diagrams of Fig. 6 (2-valve pre-amplifier) and Fig. 9 (3-valve pre-amplifier). The series input resistors should be connected directly between the input sockets and the selector switch.

It should be noted that the input-selecting switch for both circuits has a third set of controls not indicated in the circuit diagrams, which short-circuit the unwanted inputs to earth. Details of these switches are shown in Figs. 14 and 15.

Details of the low-pass filter switch SB used in the 3-valve pre-amplifier are drawn in Fig. 16. The construction of the high-pass filter switch SC is such that the capacitors of the filter are progressively connected in parallel. The wafer arrangement is drawn in Fig. 17. This arrangement reduces switch clicking and also obviates the need for large capacitors in the network.

The connection to the programme-recording output jack on the front panel of the 3-valve pre-amplifier should be made with coaxial cable from the junction of R27 and C13. This jack socket and that of the 2-valve circuit may be connected to a coaxial output socket on the back of the pre-amplifier if it is needed for a fixed recorder.

PERFORMANCE
The values for hum and noise in the pre-amplifiers which are quoted below for each input channel have been measured with each of the pre-amplifiers connected to a 20W power amplifier. The measurements were made at the output socket of the power amplifier when the input terminals of the pre-amplifier were open-circuited. The frequency-response curves were also obtained with these combinations of pre-amplifier and power amplifier.

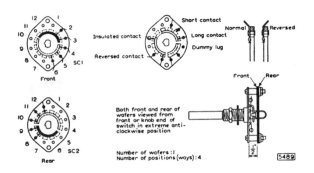

Fig. 17—Details of high-pass filter switch SC (**three-valve** pre-amplifier)

The sensitivity figures given below provide full outputs from both pre-amplifiers. Total harmonic distortion in the two-valve pre-amplifier is less than 0·15% at the rated output level and is only 0·24% at ten times this output. Total harmonic distortion in the 3-valve pre-amplifier is less than 0·1% at the rated output and only 0·65% at ten times the rated output. A rapid increase in distortion does not occur until the pre-amplifiers are considerably overloaded.

PICK-UP INPUT CHANNELS
Equalisation curves for the magnetic and the crystal pick-up input channels in both pre-amplifiers are drawn in Fig. 18. The difference in sensitivities (indicated in this figure) between the positions for microgroove and 78 r.p.m. records is achieved with the different basic levels of feedback provided at the respective positions of the selector switch. The change is necessary because standard discs are recorded at a higher level than microgroove discs.

2- AND 3-VALVE PRE-AMPLIFIERS

Magnetic Pick-up Position

	2-valve pre-amplifier	3-valve pre-amplifier
Input Impedance	100kΩ	100kΩ
Sensitivity at 1 kc/s		
(a) microgroove	4·8mV	7mV
(b) 78 r.p.m.	13mV	12mV
Hum and Noise (below 20W)		
(a) microgroove	55dB	53dB
(b) 78 r.p.m.	57dB	58dB

This channel is most suitable for pick-up heads of the variable-reluctance type, but moving-coil types which have higher outputs can be used if a larger value of series resistance is included.

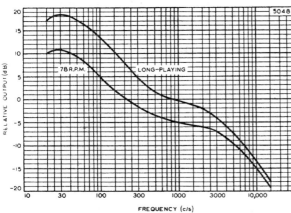

Fig. 18—*Equalisation characteristics for pick-up input channels for* **both** *circuits*

Crystal Pick-up Position

	2-valve pre-amplifier	3-valve pre-amplifier
Input Impedance	100kΩ	100kΩ
Sensitivity at 1kc/s		
(a) microgroove	70mV	150mV
(b) 78 r.p.m.	210mV	270mV
Hum and Noise (below 20W)		
(a) microgroove	55dB	53dB
(b) 78 r.p.m.	57dB	58dB

Low- and medium-output crystal pick-up heads can be used for this channel. The input is loaded with the 100kΩ resistor (causing bass loss) in order that its characteristic shall approximate to that of a magnetic cartridge, and to allow the same feedback network to be used. This produces the best compromise with most types of pick-up head.

However, if the head is not suitable for this form of loading, or if its output is too high, then it can be connected to the auxiliary input socket, which is discussed fully below. With this channel, the pick-up output is fed into a 1MΩ load which compensates automatically for the recording characteristic.

TAPE PLAYBACK INPUT CHANNEL

	2-valve pre-amplifier	3-valve pre-amplifier
Input Impedance	80kΩ (approx)	80kΩ (approx)
Sensitivity at 5kc/s	4mV	2·5mV
Hum and Noise (below 20W)	52dB	47dB

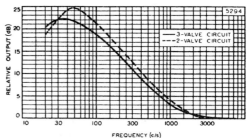

Fig. 19—*Equalisation characteristics of tape-playback input channel for* **both** *circuits*

The tape-equalisation characteristic used in each pre-amplifier is shown in Fig. 19. Each channel is intended for replaying pre-recorded tapes using high-impedance heads, and the characteristics adopted result in good performance with these heads. The frequency-response curve for playback with EMI test tape TBT1 is drawn in Fig. 20. If a greater sensitivity is required, the value of the series input resistor can be decreased until the desired sensitivity is obtained.

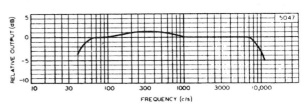

Fig. 20—*Tape-playback frequency-response characteristics using EMI test tape TBT1 for* **both** *circuits*

MICROPHONE INPUT CHANNEL
(two-valve circuit only)

Input Impedance	1MΩ
Sensitivity	7·5mV
Hum and Noise (below 20W)	44dB

The frequency-response characteristic for this channel is given in Fig. 21. The microphone input channel is intended for use with high-impedance systems such as crystal microphones or magnetic microphones with transformers.

2- AND 3-VALVE PRE-AMPLIFIERS

RADIO INPUT CHANNEL

	2-valve pre-amplifier	3-valve pre-amplifier
Input Impedance	1MΩ	1MΩ
Sensitivity	330mV	250mV
Hum and Noise (below 20W)	63dB	63dB

The frequency-response characteristics of both pre-amplifiers are given in Fig. 21. With the values of impedance and sensitivity quoted above, this channel should meet most requirements. Other values can

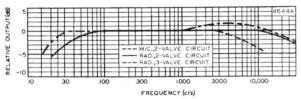

Fig. 21—Frequency-response characteristics of radio input channels for **both** circuits and of microphone channel for **two-valve** pre-amplifier

easily be obtained, however, by altering the feedback resistor and the series input resistor. If the input impedance of the channel is too high, it can be reduced by connecting a resistor of the appropriate value between the input end of the series resistor and the chassis.

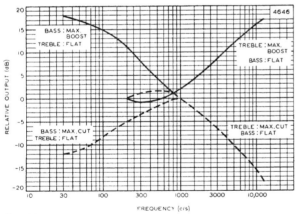

Fig. 22—Tone-control characteristics for **two-valve** pre-amplifier

AUXILIARY INPUT CHANNEL

It can be seen from the circuits that the auxiliary channel is identical with the radio input channel in both pre-amplifiers. The channels can therefore be used for high-output crystal pick-ups, for example, or for a tape pre-amplifier such as that described in Chapter 12. If it is desired to use a microphone with the 3-valve circuit, a separate microphone pre-amplifier or an input-mixing pre-amplifier can be connected to the auxiliary socket. The auxiliary input is taken to a jack socket at the front of the chassis. This makes it easier to connect a portable tape recorder. The jack-socket termination of the auxiliary input position is such that insertion of the jack disconnects the coaxial socket on the rear panel.

TONE CONTROLS

The treble and bass tone-control characteristics of the pre-amplifiers are shown in Figs. 22 and 23. These indicate that an adequate measure of control is provided in both units for most applications.

Low-impedance controls have been adopted in the 2-valve circuit so that any capacitance resulting from the use of long coaxial leads between the pre-amplifier and main amplifier will have a minimum effect on the output impedance of the pre-amplifier.

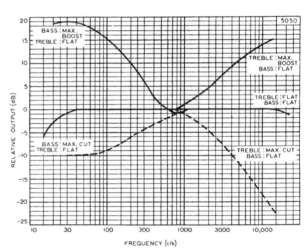

Fig. 23—Tone-control characteristics for **three-valve** pre-amplifier

AUXILIARY OUTPUT POSITION

The output from the second EF86 of both circuits is available at this auxiliary position enabling a record of programme material to be made with tape equipment. This additional output is taken to a jack socket at the front of the chassis. Excellent recordings can be made even when the input signal is derived from a low-output magnetic pick-up.

An output of about 250mV is available at this socket, and the impedance is low. Recording equipment plugged into this socket should not have an input impedance less than 500kΩ. The tone controls and the filter networks are inoperative when this output is used.

2- AND 3-VALVE PRE-AMPLIFIERS

FILTER NETWORKS (three-valve circuit only)

The characteristic of the low-pass filter has a slope of approximately 20dB per octave. The components of the network are arranged for operation at 5, 7 and 9kc/s in positions (a), (b) and (c) respectively of the switch SB in Fig. 3. Position (d) of the switch gives a flat characteristic.

The characteristic of the high-pass filter has a slope of approximately 12dB per octave. Operation is at 160, 80 and 40c/s respectively at positions (a), (b) and (c) of the switch SC. Position (d) of the switch cuts at 20c/s: this is considered preferable to allowing the response to continue to lower frequencies because of the possibility of the amplifier or loudspeaker being overloaded.

D.C. CONDITIONS

The d.c. voltages at points in the equipment should be tested with reference to Table 3 for the 2-valve circuit and to Table 4 for the 3-valve circuit. The results shown in these tables were obtained using an Avometer No. 8.

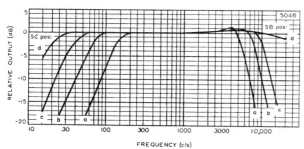

Fig. 24—Filter network characteristics for **three-valve** pre-amplifier

TABLE 3
D.C. Conditions for 2-valve Pre-amplifier

	Point of Measurement	Voltages (V)	D.C. Range of Avometer* (V)
	H.T.	230	1000
	C9	200	1000
2nd EF86	Anode	75	1000
	Screen grid	90	1000
	Cathode	2·3	10
1st EF86	Anode	70	1000
	Screen grid	50	1000
	Cathode	1·7	10

*Resistance of Avometer
1000V-range, resistance= 20MΩ
10V-range, resistance=200kΩ

TABLE 4
D.C. Conditions for 3-valve Pre-amplifier

	Point of Measurement	Voltages (V)	D.C. Range of Avometer* (V)
	H.T.	250	1000
	C22	220	1000
	C8	200	1000
ECC83	2nd Anode	200	1000
	2nd Cathode	1·8	10
	1st Anode	160	1000
	1st Cathode	1·5	10
2nd EF86	Anode	90	1000
	Screen grid	90	1000
	Cathode	1·8	10
1st EF86	Anode	90	1000
	Screen grid	75	1000
	Cathode	1·8	10

*Resistance of Avometer
1000V-range, resistance= 20MΩ
10V-range, resistance=200kΩ

CHAPTER 10
Four-channel Input-mixing Pre-amplifier

It may often be desirable to use several signal sources with an amplifying unit which has provision for only one input signal. The circuit described in this chapter is capable of handling four input signals and of supplying a mixed output voltage suitable for driving a single-input amplifier. Two of the input channels of the mixer are suitable for microphone signals, a third for a radio or equalised-tape input and a fourth for crystal pick-up signals.

The circuit as it stands provides an output voltage of 40mV and was intended to be used with the version of the 10W amplifier in which the tone-control network is disconnected, described in Chapter 6. With a simple modification to the output stage, the mixer will provide an output voltage of up to 800mV.

CIRCUIT DESCRIPTION

Both microphone input stages of the input mixing circuit are identical. Each is equipped with the Mullard low-noise pentode, type EF86, operating with grid-current bias obtained by means of the high-valued grid resistor R1.

The internal impedance of a crystal microphone is predominantly capacitive, and the capacitance is of the order of 2000pF. Therefore, to avoid loss in terminal voltage at low audio frequencies, the microphone should be connected to a high impedance input stage. If a low value of resistance of 1·5MΩ is chosen for R1, for example, the combination of the series capacitive elements of the microphone, the grid-circuit capacitance C1 and the grid resistance R1 will result in a loss of about one-third of the output voltage from the microphone at a frequency of 100c/s. Consequently, a value of 10MΩ has been selected for R1 to provide the high-impedance input for the crystal microphone and to prevent the loss of bass output voltage.

If it is required to use a low-impedance microphone such as a moving-coil or ribbon type, the mixer can be made suitable by using a step-up transformer in the grid circuit of either EF86. A switched arrangement for low- and high-impedance microphones is shown in Fig. 2. The connections marked A and B in Fig. 2 should replace those similarly marked in Fig. 1. The output leads of the microphone transformer should be made as short as possible to avoid hum pick-up and loss in treble response.

The output from each microphone input stage is RC coupled to the grid of one half of the Mullard high-μ double triode, type ECC83. The radio and pick-up input stages are also connected to this grid by way of the resistors R11 and R12, RV13. This half of the ECC83 is arranged as a voltage-amplifying stage.

Prototype of Input-mixing Pre-amplifier

INPUT-MIXING PRE-AMPLIFIER

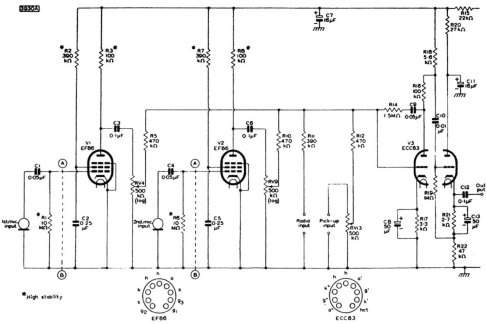

Fig. 1—Circuit diagram of input-mixing pre-amplifier

LIST OF COMPONENTS

Resistors

Circuit ref.	Value		Tolerance (±%)	Rating (W)
[1]R1, R6	10	MΩ	10	½
[1]R2, R7	390	kΩ	10	½
[1]R3, R8	100	kΩ	10	½
RV4, RV9	500	kΩ	logarithmic potentiometer	
R5, R10	470	kΩ	10	½
R11	390	kΩ	10	½
R12	470	kΩ	10	½
RV13	500	kΩ	logarithmic potentiometer	
R14	1·5	MΩ	10	½
R15	22	kΩ	10	½
[2]R16	100	kΩ	10	½
R17	3·3	kΩ	10	½
[2]R18	5·6	kΩ	10	½
R19	1	MΩ	10	½
R20	27	kΩ	10	½
R21	2·7	kΩ	10	½
R22	47	kΩ	10	½
[1]R23 (Fig.2)	10	MΩ	10	½

1. High stability, cracked carbon
2. Values may be altered for higher output voltages

Valveholders

B9A (noval) nylon-loaded, with screening skirt and flexible mounting (two for EF86s). McMurdo, XM9/UXG.1, skirt 95

B9A (noval) nylon-loaded, with screening skirt. McMurdo, XM9/AU, skirt 95

Capacitors

Circuit ref.	Value		Description	Rating (V)
C1, C4	0·05	μF	paper	250
C2, C5	0·25	μF	paper	250
C3, C6	0·1	μF	paper	250
[3]C7	16	μF	electrolytic	350
C8	50	μF	electrolytic	12
C9	0·05	μF	paper	250
C10	0·01	μF	paper	250
[3]C11	16	μF	electrolytic	350
C12	0·1	μF	paper	250
C13	50	μF	electrolytic	12
C14 (Fig. 2)	0·005	μF	paper	150

3. C7, C11: (16+16) μF wire-ended, double electrolytic capacitor

Valves

Mullard EF86 (two), ECC83

Miscellaneous

Supply input plug. Elcom, PO5
Input socket (four), coaxial
  Aerialite, 149
  Belling Lee, L.604/S/CD
Input socket (fixed). Belling Lee, L.789/CS.
Input selector switch, 1 pole, 2-way.
Output socket, coaxial
  Aerialite, 149
  Belling Lee, L.604/S/CD

The potentiometers RV4, RV9 and RV13 serve for the adjustment of signal level and the mixing of the microphone and pick-up input channels. Adjustment for the fourth channel will be achieved by means of the gain control incorporated in the radio unit used at the radio input terminals. (If it is required, control of the radio input can be achieved by connecting a potentiometer to the radio input socket in the way that RV13 is joined to the pick-up socket.)

The values of the resistors R5, R10, R11 and R12 have been chosen so that, in combination with the potentiometers RV4, RV9 and RV13, they reduce any interaction between the channels considerably. These fixed resistors also ensure that the grid of the ECC83 will not be short-circuited when any one of the potentiometers is set at a minimum.

The output stage of the mixer unit is suitable as it is shown in Fig. 1 for use with the control-less version of the 10W amplifier. The circuit has been arranged so that the sensitivity at the microphone inputs is 3mV for an output voltage of 40mV, and this is sufficient for crystal microphones. The sensitivities of the other input stages, for the same output voltage, are 230mV and 250mV for the radio and pick-up terminals respectively.

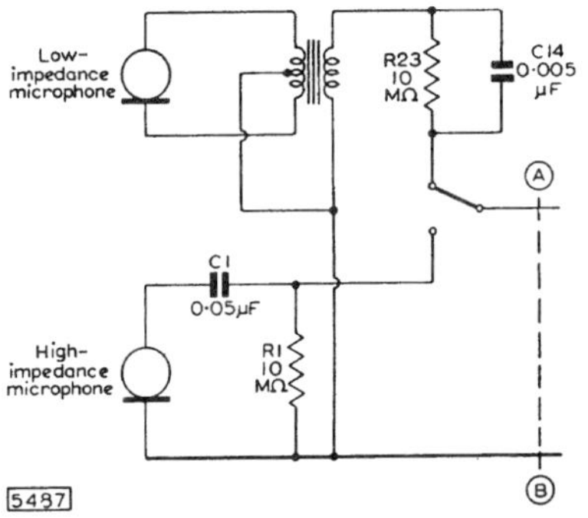

Fig. 2—Alternative switched arrangement of microphone inputs for high- and low-impedance microphones

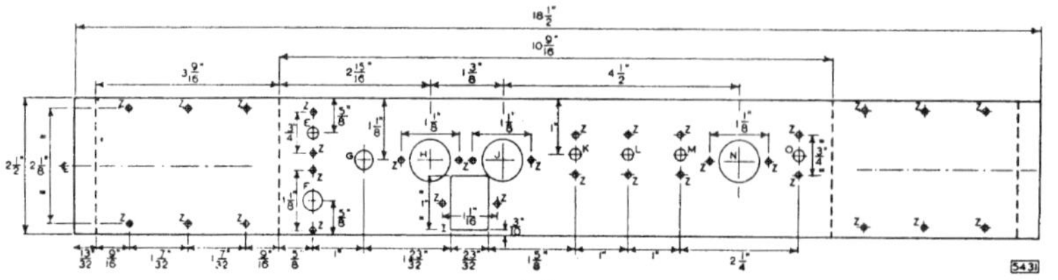

(a)

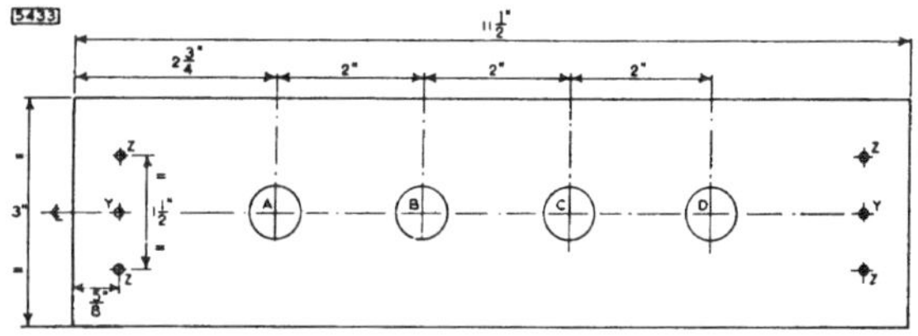

(b)

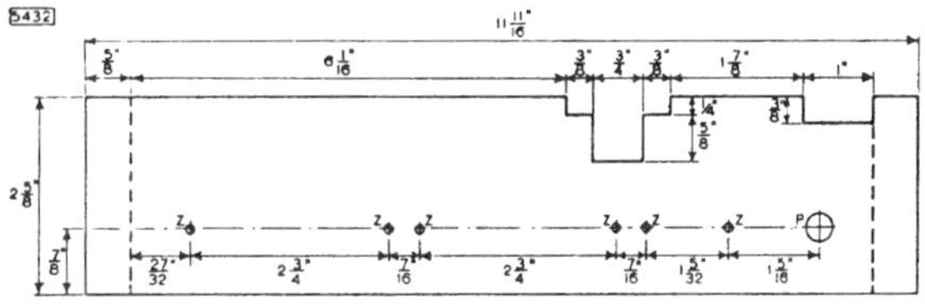

(c)

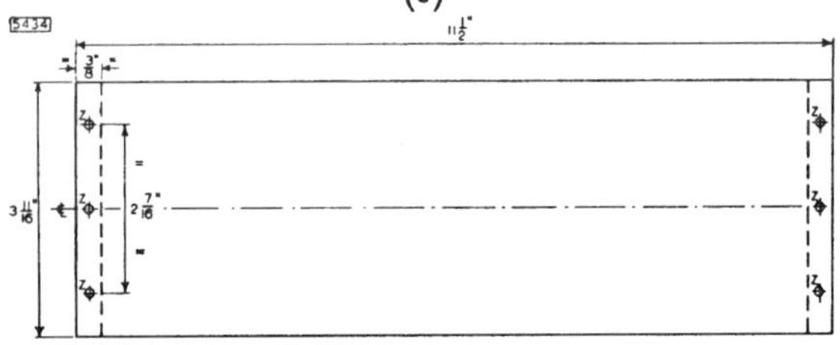

(d)

Fig. 3—Chassis details (The chassis should be bent up at 90° at all dotted lines)
(a) Rear
(b) Front
(c) Mounting plate
(d) Cover

KEY TO HOLES IN CHASSIS DRAWING

Hole	Dimension	Use	Type No.
A	$\frac{3}{8}$ in. dia.	500kΩ logarithmic potentiometer ..	—
B	$\frac{3}{8}$ in. dia.	500kΩ logarithmic potentiometer ..	—
C	$\frac{3}{8}$ in. dia.	500kΩ logarithmic potentiometer ..	—
D	$\frac{3}{8}$ in. dia.	500kΩ logarithmic potentiometer ..	—
E	$\frac{1}{4}$ in. dia.	Input socket, coaxial.	—
F	$\frac{3}{8}$ in. dia.	Input socket, fixed. Belling Lee	L.789/CS
G	$\frac{3}{8}$ in. dia.	Microphone selector switch, 1 pole, 2-way	—
H	$\frac{3}{4}$ in. dia.	B9A nylon-loaded valveholder with skirt and flexible mounting. McMurdo	XM9/UGX.1 skirt 95
I	$\frac{21}{32} \times 1\frac{5}{16}$ in.	Supply input plug, 5-pin. Elcom	P.05
J	$\frac{3}{4}$ in. dia.	B9A nylon-loaded valveholder with skirt and flexible mounting. McMurdo	XM9/UGX.1 skirt 95
K	$\frac{1}{4}$ in. dia.	Input socket, coaxial	—
L	$\frac{1}{4}$ in. dia.	Input socket, coaxial	—
M	$\frac{1}{4}$ in. dia.	Input socket, coaxial	—
N	$\frac{3}{4}$ in. dia.	B9A nylon-loaded valveholder with screening skirt. McMurdo	XM9/AU, skirt 95
O	$\frac{1}{4}$ in. dia.	Output socket, coaxial	—
P	$\frac{3}{8}$ in. dia.	Grommet hole	—
Y	Drill No. 27	4BA clearance hole	—
Z	Drill No. 34	6BA clearance hole	—

INPUT-MIXING PRE-AMPLIFIER

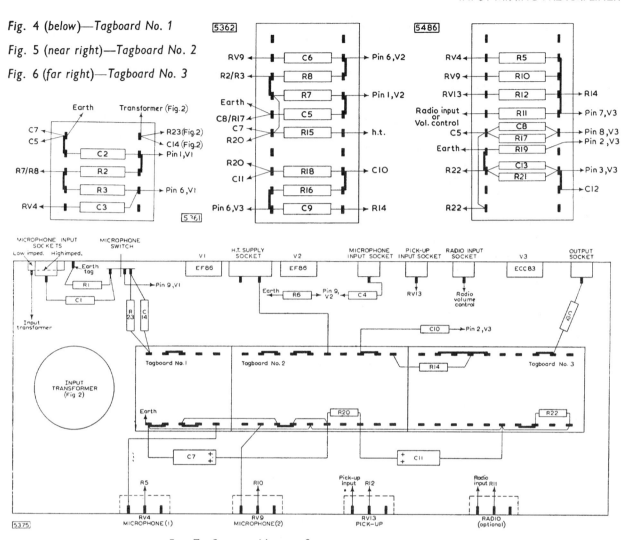

Fig. 4 (below)—Tagboard No. 1
Fig. 5 (near right)—Tagboard No. 2
Fig. 6 (far right)—Tagboard No. 3

Fig. 7—Suggested layout of components

Underside View of Prototype Input-mixing Pre-amplifier

85

INPUT-MIXING PRE-AMPLIFIER

Feedback is taken from the anode to the grid of the first half of the ECC83 by way of the components R14 and C9. The purpose of this is to provide a low impedance at the grid and hence minimise the loss in response at treble frequencies caused by Miller effect between the anode and the grid of the triode.

The output voltage is obtained from the second half of the ECC83 which has been connected as a cathode follower. This type of connection provides a low output impedance which, in Fig. 1, has the approximate value of 600Ω. Because of this low impedance, little, if any, attenuation of high notes will occur as a result of cable capacitance if a long cable is required between the mixer unit and the main amplifier. The input impedance of the main amplifier must be greater than 100kΩ to ensure that the bass frequencies will not be attenuated by the coupling capacitor C12.

CONSTRUCTION AND ASSEMBLY

Chassis details of the input-mixing pre-amplifier are given in Fig. 3. Space has been left in the chassis for the transformer (Fig. 2) necessary for converting the high-impedance microphone input into one suitable for low-impedance microphones. Space has also been provided for the optional volume control for the radio input channel.

TABLE 1
D.C. Conditions

	Point of Measurement	Voltages (V)	D.C. Range of Avometer* (V)
	H.T.	300	1000
	C7	190	1000
V3 ECC83	2nd Anode	170	1000
	2nd Cathode	28	1000
	Junction of R21, R22	26·5	1000
	1st Anode	140	1000
	1st Cathode	1·5	10
V2, V1 EF86	Anode	40	1000
	Screen grid	55	1000

*Resistance of Avometer:
  1000V-range, resistance = 20MΩ
  10V-range, resistance = 200kΩ

The chassis is made up of five separate pieces of 16 s.w.g. aluminium sheet. The dimensions (in inches) of these pieces are as follows:

(a) Front panel $\quad 11\frac{1}{2} \times 3$
(b) Rear panel $\quad 18\frac{1}{2} \times 2\frac{1}{2}$
(c) Mounting strip $\quad 11\frac{11}{16} \times 2\frac{5}{8}$
(d) Cover plate (two) $\quad 11\frac{1}{2} \times 3\frac{11}{16}$

A suggested layout for the components of the four-channel pre-amplifier is given in Fig.7. The arrangement of the components on the tagboard is shown in Figs. 4, 5 and 6. The first stage of the pre-amplifier has been modified in this layout to incorporate the facilities for low-impedance microphones. The transformer is included, and a switching arrangement allows the first stage to be used for either high- or low-impedance microphones. Also, the optional volume control for the radio input channel is shown in Fig. 7.

PERFORMANCE

Output and Sensitivity

The maximum output voltage of the mixer unit as it is drawn in Fig.1 is 40mV. This output is obtained with an input signal voltage of 3mV at either microphone socket, 230mV at the radio terminals or 250mV at the pick-up terminals.

If greater outputs from the unit are required to drive, for example, an amplifier incorporating a tone-control network, these can be achieved simply by adjusting the coupling between the anode of the first triode of the ECC83 and the grid of the second. If the capacitor C10 is joined directly to the anode of the first triode, an output of 800mV will be available. Intermediate values of output voltage can be achieved by altering the values of R16 and R18. If, for example, R16 and R18 were each 47kΩ, the output voltage of the unit would be about 400mV.

If the required output voltage from Fig.1 had been obtained by attenuating the output voltage from the cathode load in the final stage, the low output impedance resulting from the cathode-follower action would have been lost.

Frequency Response

The response curve of the mixer unit, measured between the microphone socket and the output terminals is shown in Fig. 8. The curve is flat to within ±3dB, relative to the response level at 1kc/s, from 20c/s to 20kc/s. The response curve measured between either the radio or the pick-up socket and the output terminals is shown also in Fig. 8. Because the EF86 stages are not included for this second position of measurement, the bass response is slightly extended and the curve is flat to within ±2dB, compared with the 1kc/s level, from 15c/s to 20kc/s.

Hum and Noise

Measurements of hum and noise were made with 100kΩ resistors connected across the microphone and pick-up terminals and with the potentiometers fully advanced. These arrangements simulate a reasonable practical condition. The mixer was connected to a 10W amplifier and measurements were made across a 15Ω load resistor. The voltage measured in this way was 38mV.

The full output of the amplifier is 10W which corresponds to a voltage of 12·3V across the 15Ω load resistor. Consequently the hum and noise level is 50dB below 10W. The background level of the power amplifier alone is better than 70dB below 10W, so the figure of 50dB must be attributed to hum and noise in the mixer unit.

D.C. CONDITIONS

The d.c. voltages at points in the equipment should be tested with reference to Table 1. The results shown in this table were obtained with an Avometer No. 8.

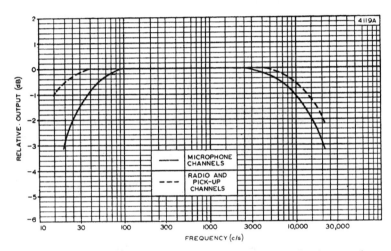

Fig. 8—Frequency-response curves for microphone inputs and radio or pick-up signals

CHAPTER 11
Three-watt Tape Amplifier

The functions associated with both the recording and playback processes are, in most tape recording equipment, fulfilled by a single amplifier with suitable switching facilities. It is only with a few of the more elaborate and costly items of equipment that separate recording and playback amplifiers are used.

The circuit to be described in this chapter combines the requirements of both processes, and it can be used with many of the tape decks that are fitted with a combined record-playback head and a separate erase head. With such a deck, the amplifier forms a self-contained recording and reproducing unit which is capable of an excellent performance, whether it is reproducing recordings made on the same unit, or whether pre-recorded tapes are used.

Any tape amplifier must, if an acceptable performance is to be achieved, provide compensation for the unequal response over the audio-frequency range that is inherent in the process of magnetic recording. In keeping with general practice, the treble equalisation is associated with the recording channel and the bass equalisation is incorporated during playback.

It has been pointed out in Chapter 2 that the extent of treble attenuation is dependent on the tape speed.

It is thus desirable to provide separate high-frequency correction for the different tape speeds. In this amplifier, equalisation is provided for speeds of $3\frac{3}{4}$, $7\frac{1}{2}$ and 15 inches per second.

In addition to the basic equalisation, some measure of tone control may be desired in a complete recorder. Equalisation is arranged to provide a level overall frequency characteristic. The tone control provided allows for modifications to be made to the playback characteristic so that the treble response can be attenuated to suit the demands of the individual listener. The control gives 18dB of cut at 10kc/s.

The quality of the performance of the complete equipment is necessarily limited by the output stage of the playback system. A higher standard of reproduction can be achieved if a good-quality amplifier and pre-amplifier are fed from the low-level output of the playback amplifier.

Excluding the limitations mentioned in the previous paragraph, and the distortion introduced by the tape itself, the major source of distortion in this unit occurs in the recording stage. The operating conditions for the second section of the ECC83 are chosen as a compromise between the requirements of gain and output voltage and the desirability of maintaining

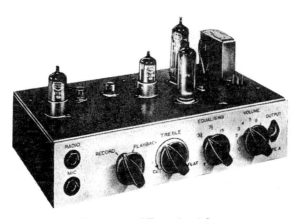

Prototype of Tape Amplifier

a high a.c./d.c. load ratio. The voltage required from this stage to provide a recording current of 200µA depends on the series resistance between the anode and the recording head. A low value of this resistance, whilst reducing the output voltage requirements, also increases the distortion of this stage. The total harmonic distortion in the recording stage at the peak level of 20V which has been adopted in this design is not more than 2% at 1kc/s.

CIRCUIT DESCRIPTION

The circuit diagram of the combined record-playback amplifier is given in Fig. 1.

Three stages of the circuit are common to both recording and playback processes. A fourth stage acts as an r.f. oscillator for the biasing and erasing signals when recording, and is used as a power-output stage in the playback process. A subsidiary stage of the recording amplifier, which is excluded from the playback circuit, is the recording-level indicator stage.

Controls

Four controls are provided in the circuit:

(1) The switch SA sets the amplifier for either the recording or playback condition.
(2) The switch SB gives the appropriate equalisation for a tape speed of either 15, $7\frac{1}{2}$ or $3\frac{3}{4}$ inches per second. The switch must be in the correct position during playback as well as when recording.
(3) The tone control RV15, which is operative only during playback, is situated in the anode circuit of the first section of the ECC83, and gives variable treble cut.
(4) The gain control RV20 operates during both recording and playback processes. It does not influence the low-level output which is available at the anode of the second stage of the amplifier.

Valve Complement

The amplifier uses four Mullard valves and one Mullard germanium diode. These are:

(a) Type EF86, low-noise pentode, used in the input stage for both recording and playback functions.
(b) Type ECC83, double triode. The first section of the valve is used in the equaliser stage for recording and playback. The second section is used as the output stage when recording and as a voltage-amplifying stage during playback.
(c) Type EL84, output pentode. When recording, this is used as the oscillator valve, and in the playback process, it is used in a single-ended power-output stage to drive the loudspeaker.
(d) Type EM81, tuning indicator, used in the recording-level stage.
(e) Type OA81, germanium diode, used as the indicator-circuit rectifier.

Input Stage

The pentode, type EF86, acts as a voltage amplifier for both recording and playback processes. It is possible to record from either microphone or radio sources. Both inputs are fed to the grid of the valve, the radio being attenuated to the level of the microphone input. The switching is achieved by inserting the jacks so that only one input may be used at a time.

Equaliser Stage

One section of the double triode, type ECC83, is used in the second or equaliser stage of both processes. The tone control which is operative only during playback, is also located in this stage.

During the recording process, a resonant circuit, containing a Ferroxcube pot-core inductor L1 is used to provide treble equalisation. The value of tuning capacitance in the resonant circuit is selected by the equalisation switch SB to give maximum treble boost appropriate to the particular tape speed. The degree of boost is controlled by the resistor R7 and the damping resistors R8 and R9 connected in parallel at SB2 with the capacitors C7 and C8. The steep fall in boost that occurs below the resonant frequency is modified by damping the inductor and by partially shunting R7 at SB1 at the appropriate frequencies. The treble boost obtained will be correct for most combinations of tape and head, but it may be too great for some. If this is so, the damping on L1 should be increased by connecting a resistor in parallel with C9 and reducing the values of R8 and R9. The optimum values should be determined by listening tests.

During playback, the correct feedback for bass equalisation is provided by the resistors R11, R12 and R13 arranged on SB3.

A low-level output of 250mV, having a source impedance of 50kΩ, can be taken from the anode load of this stage of the amplifier and can be used either during recording for monitoring purposes, or during playback for feeding an external pre-amplifier and power amplifier.

Recording Output Stage

The output from the anode of the equaliser stage is taken to the grid of the second section of the double triode by way of the gain control RV20. Further

3W TAPE AMPLIFIER

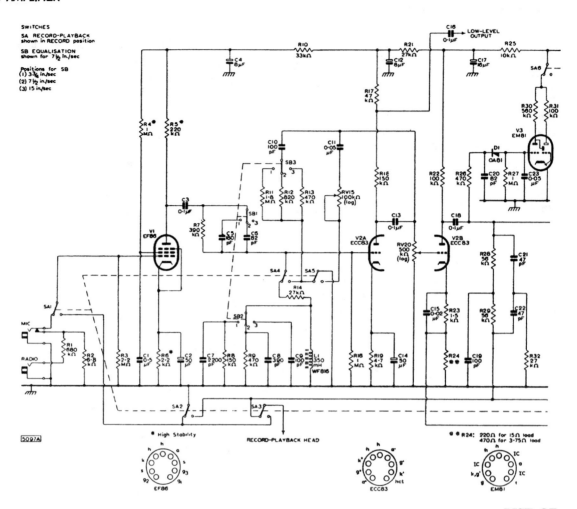

LIST OF Resistors

Circuit ref.	Value	Tolerance (±%)	Rating (W)	Circuit ref.	Value	Tolerance (±%)	Rating (W)
R1	680 kΩ	10	¼	R22	100 kΩ	10	¼
R2	6·8 kΩ	10	¼	R23	1·5 kΩ	10	¼
R3	2·2 MΩ	10	¼	R24			
¹R4	1 MΩ	5	½	for 3·75Ω speaker 470 Ω		10	¼
¹R5	220 kΩ	5	½	for 15Ω speaker 220 Ω		10	¼
¹R6	2·2 kΩ	5	½	R25	10 kΩ	10	¼
R7	390 kΩ	10	¼	R26	470 kΩ	10	¼
R8	150 kΩ	10	¼	R27	1 MΩ	10	¼
R9	470 kΩ	10	¼	R28	56 kΩ	10	¼
R10	33 kΩ	10	¼	R29	56 kΩ	10	¼
R11	1·8 MΩ	10	¼	R30	560 kΩ	10	¼
R12	820 kΩ	10	¼	R31	100 kΩ	10	¼
R13	470 kΩ	10	¼	R32	27 kΩ	10	¼
R14	27 kΩ	10	¼	R33	22 kΩ	10	¼
RV15	100 kΩ logarithmic potentiometer			R34	18 kΩ	10	¼
R16	1 MΩ	10	¼	R35	680 kΩ	10	¼
R17	47 kΩ	10	¼	R36	6·8 kΩ	10	1
R18	150 kΩ	10	¼	R37	1 kΩ	10	¼
R19	4·7 kΩ	10	¼	R38	150 Ω	10	1
RV20	500 kΩ logarithmic potentiometer			R39	1 kΩ	10	¼
R21	27 kΩ	10	¼	R40	2·2 kΩ	10	¼

¹High stability, cracked carbon

3W TAPE AMPLIFIER

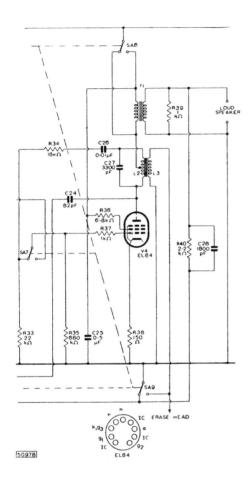

Fig. 1—Circuit diagram of self-contained tape amplifier

Capacitors

Circuit ref.	Value		Description	Rating (V)
C1	0·5	μF	paper	350
C2	50	μF	electrolytic	12
C3	0·1	μF	paper	350
C4	8	μF	electrolytic	350
C5	180	pF	silvered mica	
C6	82	pF	silvered mica	
C7	2200	pF	silvered mica	
C8	390	pF	silvered mica	
C9	100	pF	silvered mica	
C10	100	pF	silvered mica	
C11	0·05	μF	paper	150
C12	8	μF	electrolytic	350
C13	0·1	μF	paper	350
C14	50	μF	electrolytic	12
C15	0·02	μF	paper	350
C16	0·1	μF	paper	350
C17	16	μF	electrolytic	350
C18	0·1	μF	paper	350
C19	100	pF	silvered mica	
C20	82	pF	silvered mica	
C21	47	pF	silvered mica	
C22	47	pF	silvered mica	
C23	0·05	μF	paper	150
C24	82	pF	silvered mica	
C25	0·5	μF	paper	350
C26	0·01	μF	paper	350
C27	3300	pF	silvered mica	
C28	1800	pF	silvered mica	

Tolerance of all silvered mica capacitors is ±10%

Valves and Germanium Diode

V1	Low noise pentode,	Mullard type EF86
V2	Double triode,	Mullard type ECC83
V3	Tuning indicator,	Mullard type EM81
V4	Output pentode,	Mullard type EL84
D1	Germanium diode,	Mullard type OA81

Valveholders

B9A (noval) (two). McMurdo BM9/U
B9A (noval) nylon-loaded with screening skirt (two). McMurdo, XM9/AU, skirt 95

Output Transformer

Primary Impedance, 5kΩ
Secondary Impedance, 3·75 and 15Ω

COMPONENTS

Miscellaneous

Loudspeaker sockets (red and black). Belling Lee, L.316
Output socket recessed coaxial. Belling Lee, L.734/S
Record-playback head coaxial socket. Belling Lee, L.604/S
Erase head coaxial socket. Belling Lee, L.604/S
Supply input socket. Elcom, P04
Input jack (radio). Igranic, P71
Input jack (microphone). Igranic, P72
Record-playback switch SA.
 Shirley Laboratories, 16370/B3
 Specialist Switches, SS/567/A
Equalisation switch SB, 3-pole, 3-way.
(Note: Details of proprietary switches may not be identical with those given in the diagrams.)

Inductors

Equalisation Coil L1 (350mH)
 Wound Ferroxcube pot core, Mullard type WF816
Oscillator Coil L2, L3
 Former: Standard Aladdin
  ½in. dia.; ⅜in. winding length
 Slug: ⅜in. dia.; ½in. long, centred on winding
 Primary: 400 turns of 38 s.w.g. silk-covered copper wire tapped at 360 turns from anode end
 Secondary: 50 turns of 34 s.w.g. silk-covered copper wire

Commercial Components

Manufacturer	Type No.
Colne	03077
Elden	1114
Elstone	OT/3
Gilson	W.O.767
Hinchley	1534
Parmeko	P2641
Partridge	P4073
Wynall	W1452

3W TAPE AMPLIFIER

high-frequency boost is added to the recording signal by the capacitor C15 in combination with the resistor R23.

The recording signal from the anode of the second section of the ECC83 is taken by way of a parallel-T network to the recording head. The network presents its highest impedance at the biasing frequency. Bias is fed to the recording head immediately after the T-network. This arrangement produces a substantially constant current drive to the recording head and provides efficient rejection of the bias voltage at the anode of the output valve.

H.F. Oscillator (Record) or Power Output Stage (Playback)

The output pentode, type EL84, acts as an audio output stage during playback. In the recording process, the EL84 is used to provide the h.f. oscillations for the biasing and erasing signals.

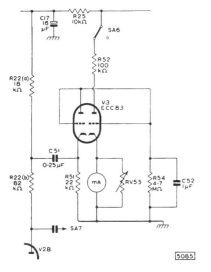

Fig. 2—Circuit for meter-indication of recording level

The bias signal is introduced into the recording head through the capacitor C24, the value of which determines the bias current flowing in the head. The bias voltage is obtained from the anode of the EL84.

The oscillator coil and the choice of component values for the oscillator circuit will depend on the types of record-playback and erase heads used. The details given with Fig. 1 are those suitable for record-playback heads having an impedance between 15 and 30kΩ and erase heads with an impedance between 200 and 300Ω. Details for heads with other values of impedance should be obtained from the manufacturer of the tape deck used.

The bias oscillator coil and the primary winding of the output transformer are arranged in series. The latter is by-passed by the switch SA8 when in the recording position.

The presence of the capacitor C25 prevents an abrupt cessation of the oscillations when the amplifier is switched from the recording to the playback position, and thus prevents magnetisation of the record-playback head. The erase head is earthed for the playback process by the switch SA9.

On playback, approximately 10dB of negative feedback is used, the feedback voltage being taken from the secondary winding of the output transformer to the cathode of the second section of the double triode. The harmonic distortion in the output stage is not more than 3% at 1kc/s for an output level of 3 watts. The output power from the playback amplifier is fed by way of the transformer T1 to either a 3·75Ω or a 15Ω speaker.

Recording-level Indicator

A tuning indicator, type EM81, is fed from the anode of the second section of the ECC83 through a detector circuit using a germanium diode, type OA81.

The value of the resistor R31 in the target-anode circuit governs the sensitivity of the indicator, and has been chosen to give a sufficiently high sensitivity to allow a large series resistance R26 to be used between the diode and the second anode of the ECC83. This large resistance minimises the loading effect on the recording output stage.

The operating conditions of the EM81 are normally chosen so that the target shadow 'closes' for a recording current of 200μA. They can however be chosen so that the shadow 'closes' at lower peak recording levels if reduced peak distortion is desired at the expense of the signal-to-noise ratio.

In the playback position, switch SA6 disconnects the recording-level indicator stage from the h.t. supply, and this gives a positive indication of the position of the record-playback switch SA.

If a meter indication of the recording level is required, the EM81 stage of Fig. 1 can be replaced by the circuit given in Fig. 2. The valve used is an ECC83. The signal is taken from a point in the anode load R22 of the recording output stage (V2B in Fig. 1), and the milliammeter (1mA, full-scale deflection) is included in the cathode circuit of the second half of the ECC83 of Fig. 2. The variable resistor RV53 should be chosen to suit the resistance of the milliammeter.

3W TAPE AMPLIFIER

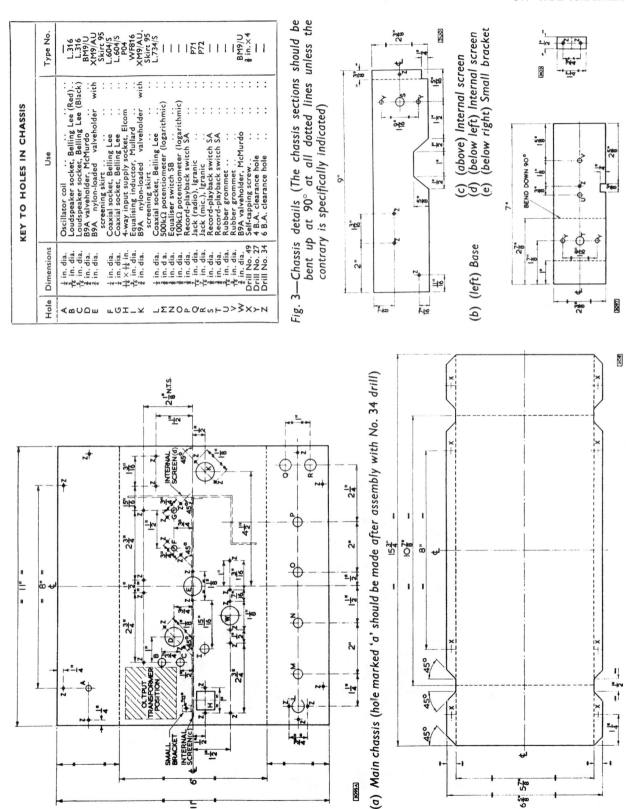

Fig. 3—*Chassis details (The chassis sections should be bent up at 90° at all dotted lines unless the contrary is specifically indicated)*

(a) Main chassis (hole marked 'a' should be made after assembly with No. 34 drill)
(b) (left) Base
(c) (above) Internal screen
(d) (below left) Internal screen
(e) (below right) Small bracket

3W TAPE AMPLIFIER

CONSTRUCTION AND ASSEMBLY

The chassis is made up of five separate pieces of 16 s.w.g. aluminium sheet. The dimensions (in inches) of these are:

(a)	Main chassis	11×11
(b)	Base	$15\frac{3}{4} \times 6\frac{5}{8}$
(c)	Internal screen	$9 \times 2\frac{3}{8}$
(d)	Internal screen	$7 \times 2\frac{3}{8}$
(e)	Small bracket	$1\frac{1}{4} \times \frac{1}{2}$

Each piece should be marked as shown in the chassis drawings of Fig. 3, and the holes should be cut as indicated. It is important that, when bending the sheet, the scribed lines should be exactly along the angles. This ensures that the pieces will fit together properly when assembled.

Before assembling the record-playback switch SA around the screens, it will probably be found convenient to fix the following components to the chassis:

(1) The erase and playback sockets which have to be fitted to the chassis beneath wafer 3 of switch SA.

(2) The nylon-loaded valveholder for the ECC83, complete with the skirt for the screening can. Pins 1 and 9 should face towards the coaxial sockets.

(3) A five-way tagboard. This should be bolted to the internal screen c in the position marked 'Tagboard No. 6' in the layout diagram of Fig. 11.

(4) The small fixing bracket f, which should be bolted to the screen c as indicated in Fig. 3(a).

The construction is best continued by assembling wafers 1 and 2 of switch SA around the internal screen d. The wafers should be arranged so that positions 6 and 7 are nearest the chassis and the face of each wafer described as the 'rear' in the switch diagram (Fig. 4)

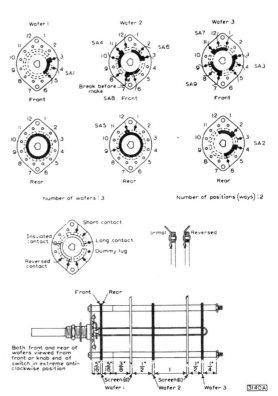

Fig. 4—Record-playback switch details

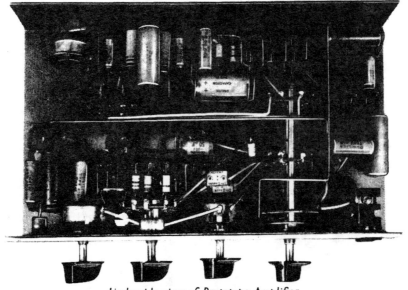

Underside view of Prototype Amplifier

3W TAPE AMPLIFIER

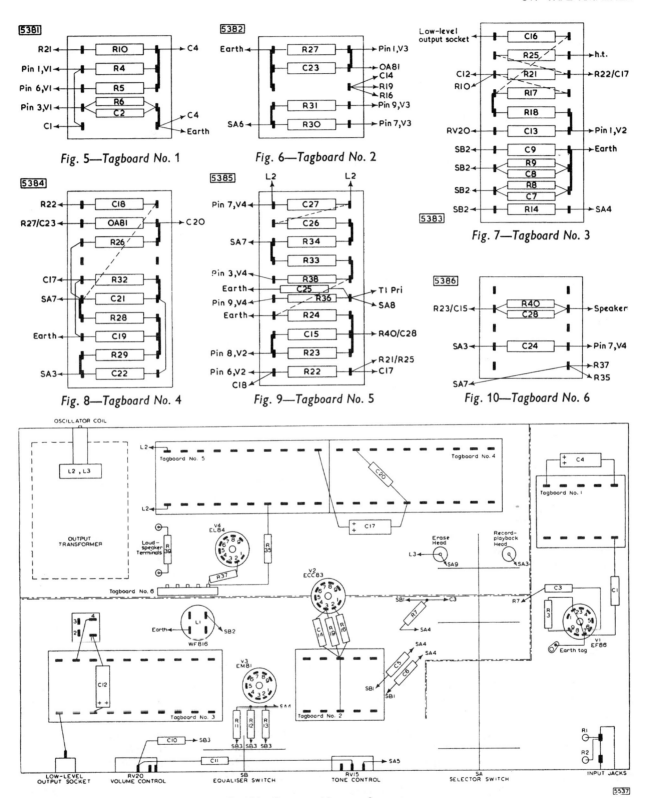

Fig. 5—Tagboard No. 1
Fig. 6—Tagboard No. 2
Fig. 7—Tagboard No. 3
Fig. 8—Tagboard No. 4
Fig. 9—Tagboard No. 5
Fig. 10—Tagboard No. 6

Fig. 11—Suggested layout of components

3W TAPE AMPLIFIER

is farthest from the switch plate. The internal screen c should be added to the assembly, both screens should be bolted together, and wafer 3 should be fitted in position, again with its 'rear' face (Fig. 4) farthest from the switch plate. The general arrangement of the switch wafers and internal screens is shown in Fig. 11. Details of spacers required for the assembly are shown in Fig. 4. The use of a shock-proof washer is recommended between the switch plate and the front panel when the switch and screens are fitted to the main chassis. A suitable layout of the other components is shown in Fig. 11. The arrangements of the small components on tagboards are shown in Figs. 5 to 10. Details of the equaliser switch are given in Fig. 12.

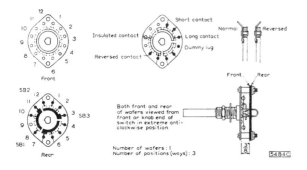

Fig. 12—Equaliser switch details

PERFORMANCE

Frequency Response

Treble boost is incorporated during recording and bass boost during playback. Separate equalisation is provided for tape speeds of 15, 7½ and 3¾ inches per second to give the following attainable overall response (relative to the level at 1kc/s):

15 in./sec   ±3dB from 35c/s to 16kc/s
7½ in./sec   ±3dB from 50c/s to 13kc/s
3¾ in./sec   ±3dB from 50c/s to 6kc/s

The overall response at the higher frequencies depends on the type of head used and on the magnitude of the bias current. The response figures given above and the curves drawn in Fig. 13 will normally be obtained with a bias current of 0·5 to 1·0mA through heads of medium impedance.

The playback characteristic of the amplifier at a tape speed of 7½ inches per second is designed to the C.C.I.R. specification, thus permitting excellent reproduction of pre-recorded tapes. The recording characteristic is arranged to give a flat frequency response in conjunction with this replay characteristic.

Sensitivity

The sensitivity of each amplifier is measured with the control RV20 set for maximum gain. This, of course, does not apply to the low-level output measurements: the control is not effective until after this point of the circuit.

Recording Sensitivity

(measured at 1kc/s, with recording-head audio current of 200μA)

(a) Microphone input: 2·5mV for peak
    (impedance = 2MΩ)   recording level
(b) Radio input: 250mV for peak
    (impedance = 680kΩ) recording level

Playback Sensitivity

(measured at 5kc/s for all tape speeds for 3W power output or 250mV low-level output)

(a) 15 in./sec: 5·2mV
(b) 7½ in./sec: 2·8mV
(c) 3¾ in./sec: 1·1mV

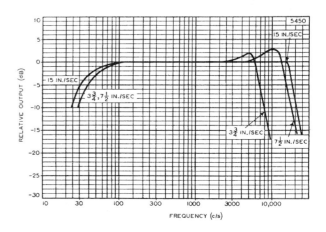

Fig. 13—Frequency-response characteristics

TEST PROCEDURE

The four tests outlined below are intended as simple, yet quite effective, checks for the combined record-playback amplifier.

The values given in the various tables and figures were obtained from the prototype amplifier, using Brenell record-playback and erase heads. The bias current used throughout was 1·0mA at a frequency of 60kc/s, and the erase-head voltage was about 25V, again at a frequency of 60kc/s.

3W TAPE AMPLIFIER

Test I—D C. Voltages

The d.c. voltages at points in the equipment should be tested with reference to Table 1. The results shown in this table were obtained using an Avometer No. 8.

Test II—Amplifier on Playback

Three pieces of equipment are required for this test:

(1) A signal generator covering a frequency range from 20c/s to 20kc/s;

(2) A valve voltmeter covering a frequency range from 20c/s to 20kc/s;

(3) A load resistor of 15Ω with a 6W rating.

The 15Ω resistor should be connected to the speaker sockets. The record-playback switch SA should be in the playback position and the tone control RV15 should be set for the flat response.

A signal from the generator, having a frequency of 5kc/s, should be applied to the record-playback socket (which normally accommodates the connection plug from the record-playback head). The consequent output signals should be measured on the voltmeter,

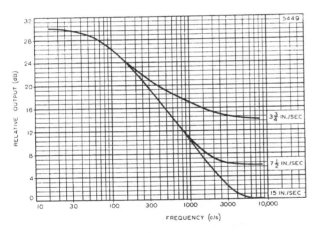

Fig. 14—Bass-boost characteristics

both at the low-level output socket and across the load resistor.

The input voltage should be adjusted to give an output voltage across the load resistor of 6·7V for each tape speed, and the input required for this output should be noted. The voltage readings that should be obtained are given in Table 2.

With the switch SB set to 15 inches per second, and the conditions obtaining in the previous paragraph, the gain control should be varied until the output voltage across the load resistor drops to 50mV. The

TABLE 1

D.C. Conditions

Point of Measurement		Voltages (V)		D.C. Range of Avometer* (V)
		(a) SA in Record position	(b) SA in Playback position	
	C17	275	271	1000
	C12	240	240	1000
	C4	218	218	1000
EL84	Anode	300	260	1000
	Screen grid	240	260	1000
	Cathode	7·5	8·0	100
EM81	Anode	60	0	1000
	Target	250	0	1000
ECC83	1st anode	160	160	1000
	1st cathode	1·5	1·5	10
	2nd anode	170	170	1000
	2nd cathode	1·5	1·5	10
EF86	Anode	65	65	1000
	Screen grid	80	80	1000
	Cathode	1·8	1·8	10

*Resistance of Avometer:
  1000V-range, resistance = 20MΩ;
  100V-range, resistance = 2MΩ;
  10V-range, resistance = 200kΩ.

TABLE 2

Playback Sensitivity

Tape Speed (in./sec)	Input (mV)	Output Voltages	
		Low level (mV)	15Ω load (V)
15	5·2	250	6·7
7½	2·8	250	6·7
3¾	1·1	250	6·7

3W TAPE AMPLIFIER

frequency of the signal should then be reduced to 100c/s and the values of boost given in Table 3 should be observed at the 15Ω-load output. The switch SB should be changed to 7½ and 3¾ inches per second, and the boost measurement should again be made.

The bass-boost characteristics for the three tape speeds are shown in Fig. 14.

The treble cut introduced by full application of the tone control RV15 should be determined for all tape speeds, and the curve obtained should correspond to that shown in Fig. 15.

Test III—Amplifier on Record

The instruments required for this test are:

(1) A signal generator covering a frequency range from 20c/s to 20kc/s;
(2) A valve voltmeter* covering a frequency range from 20c/s to 20kc/s.

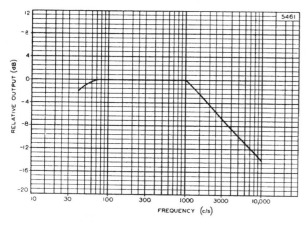

Fig. 15—Tone-control characteristic

The record-playback and erase heads should be connected to the appropriate sockets in the amplifier, and the equipment should be switched to the recording condition.

For a tape speed of 15 inches per second, a signal at 1kc/s should be applied from the generator to the radio input socket. The magnitude of this signal should be such that an output of 30mV is obtained at the low-level output socket.

When the signal frequency is switched to 15kc/s, the boost indicated in Table 4 should be observed.

---

*For accurate results two separate pieces of p.v.c. covered wire are recommended for the connection to the valve voltmeter. A coaxial cable may result in considerable errors in the measurements because of the parallel capacitance which is introduced.

With tape speeds of 7½ and 3¾ inches per second, the input signal at 1kc/s should again be adjusted to give a low-level output of 30mV. The change to input frequencies of 10 and 5kc/s should give the treble boost indicated in Table 4.

The treble-boost characteristics for the three tape speeds are shown in Fig. 15.

TABLE 3

Bass Boost

Signal frequency = 40c/s

(Output voltage for 5kc/s = 50mV)

Tape Speed (in./sec)	Voltmeter reading (V)	Output boost (dB)
15	1·18	28
7½	0·65	22
3¾	0·25	15

TABLE 4

Treble Boost

(Output voltage for 1kc/s = 30mV)

Tape Speed (in./sec)	Signal frequency (kc/s)	Voltmeter reading (mV)	Output boost (dB)
15	15	90	10
7½	11	180	15
3¾	5·5	180	15

TABLE 5

Recording Sensitivity

Signal frequency ..	1	kc/s
Tape speed	15, 7½ or 3¾	in./sec
Voltage at second anode of ECC83	20	V
Microphone input	2·5	mV
Radio input	250	mV

3W TAPE AMPLIFIER

Values for the recording sensitivity for an output voltage measured at the second anode of the ECC83 are given in Table 5. A test of the recording-level indicator should show that the EM81 'closes' with approximately 20V at this anode.

An alternative method of checking the recording amplifier is possible. For any of the tape speeds, the voltage developed across a 50Ω-resistor connected in series with the recording head can be observed for the full range of signal frequencies. The response curve so obtained should agree with the appropriate curve for the prototype amplifier, plotted in Fig. 13. For these observations, it will be necessary to disconnect one end of the resistor R34, otherwise only the bias signal will be measured.

Test IV—Bias Level

For this test, two pieces of equipment are required:

(1) A valve voltmeter which will indicate accurately at frequencies of up to 70kc/s;
(2) A resistor of 50Ω.

The resistor should be soldered in series with the earthy end of the record-playback head, and the voltage developed across the resistor, with no input signal, should be measured with the voltmeter.

The voltage developed across the resistor should be 50mV, which corresponds to a bias current of 1·0mA flowing in the 50Ω-resistor.

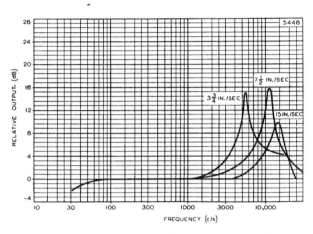

Fig. 16—Treble-boost characteristics

Power-supply Unit

CIRCUIT DESCRIPTION

The circuit diagram for a power supply suitable for use with the 3W tape amplifier (and the tape pre-amplifier described in the following chapter) is given in Fig. 17. The requirements of the unit are that it should provide (i) a direct voltage of 300V at a current of 50mA and (ii) an alternating voltage of 6·3V at a current of 2A.

The choice of rectifier will depend on the tape deck used. Normally, the Mullard full-wave rectifier, type EZ80, will be suitable. However, with tape decks that use electrical braking for the tape transport system, it is essential that the Mullard type EZ81 be used, so that the current which is required for the short braking periods can be supplied.

If the EZ80 is used, the series resistance in each anode circuit of the rectifier must be at least 215Ω; if the EZ81 is used, the minimum series resistance is 200Ω for each anode. Very few transformers meeting the specification given below will have a total winding resistance less than these minimum requirements, but should it be lower, a series resistance large enough

Prototype of Power Supply

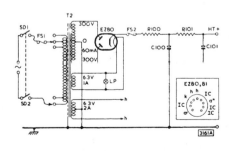

Fig. 17—Power-supply circuit

3W TAPE AMPLIFIER

LIST OF COMPONENTS

C100, C101
50+50µF electrolytic capacitor
Working voltage rating = 350Vd.c.
Min. ripple-current rating = 100mA

R100
Value to give less than 350V across C100
(Not required in prototype)

R101
Value to give 300V across C101
(820Ω, 3W, wire-wound resistor in prototype)

Valve
Full-wave rectifier, Mullard, EZ80 (EZ81)

Valveholder
B9A (noval). McMurdo BM9/U

Mains Transformer
Primary: 10–0–200–220–240
Secondaries: 300–0–300V, 60mA
3·15–0–3·15V, 2A
0–6·3V, 1A

Commercial Components

Manufacturer	Type No.
Colne	03080
Elden	1125
Elston	MT/3M
Gilson	W.O.839
Hinchley	1446
Parmeko	P2631
Partridge	H300/60
Wynall	W1547

Miscellaneous
Input securing clip. Hellerman P clip, 3180/5B
Output securing clip. Hellerman P clip, 3180/3B
Mains switch, 2-way. Bulgin, S300
Fused voltage selector. Clix, VSP393/0, P62/1
Fuseholder, Minifuse. Belling Lee, L.575
Fuse. Belling Lee Minifuse, 1A
Fuse. Belling Lee, 200mA
Lampholder (optional). Bulgin, D180/Red
Pilot lamp (optional). Bulgin, 6·3V, 300mA

Fig. 18 (right)—Chassis details of power supply

KEY TO HOLES IN CHASSIS

Hole	Dimensions	Use	Type No.
A	⅝ in. dia.	B9A valveholder, McMurdo	BM9/U
B	1⅛ in. dia.	50+50µF electrolytic capacitor	—
E	—	Mains selector switch, Clix	VSP393/0
F	½ in. dia.	Mains switch, Bulgin 2-way	S300
G	7/32 in. dia.	Minifuse holder, Belling Lee	L.575
H	⅝ in. dia.	Lampholder, Bulgin	D180/Red
X	Drill No. 12	No. 6 wood screw	—
Y	Drill No. 27	4 B.A. clearance hole	—
Z	Drill No. 34	6 B.A. clearance hole	—

to make up the minimum should be added to each anode circuit. The required value of this series resistance is derived in the way shown on page 28.

The values of the dropper resistors R100 and R101 shown in Fig. 17 should be chosen to give respectively a potential of less than 350V across the reservoir capacitor C100 and a potential of 300V across C101. It may be found that R100 is not needed.

CONSTRUCTIONAL DETAILS

The chassis consists of one piece of 16 s.w.g. aluminium sheet, 8¾ in. long, and 7 in. wide. It should be marked as shown in the chassis drawing of Fig. 18 and the holes should be cut as indicated. Mounting holes for the mains switch, the fused voltage-selector and the pilot lamp are shown in the figure, but if it is so desired, these components may be mounted elsewhere when, of course, there will be no need to cut the particular holes in this chassis.

The chassis drawing shown in Fig. 18 is for a mains transformer of the inverted-mounting type. If a different type is used, it will be necessary to drill grommet holes to enable the leads to be taken through the chassis.

The assembly of the power unit should be accomplished quite easily by referring to the layout diagram of Fig. 19. This figure again caters for a transformer of the inverted-mounting type.

To avoid unnecessary expense, input and output plugs have not been used: securing clips suffice to

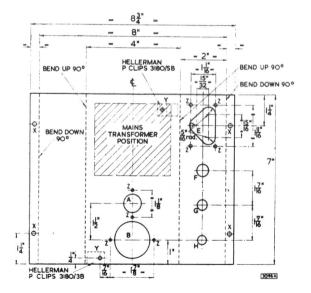

anchor the mains and h.t. leads. The pilot lamp is also an optional component.

It is very important when wiring the electrolytic capacitor to ensure that the correct section is used as the reservoir capacitor C100. The section which is identified as the 'outer', or else marked with a red spot, should be used for this component.

Underside View of Power Supply

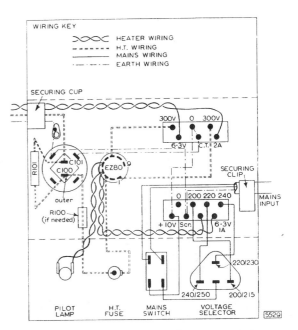

Fig. 19—Suggested layout for power supply

CHAPTER 12
Tape Pre-amplifier

The tape-recording pre-amplifier described in this chapter is intended to be used in conjunction with a high-quality pre-amplifier and power amplifier. The unit combines the function of both recording and playback amplification, although, for the playback operation, it acts only as an equalising stage, giving sufficient output voltage to drive the high-quality system.

The general principle of the design has been to preserve simplicity as far as is compatible with high-quality performance. The distortion introduced in the recording channel has been reduced to such an extent that is is probable that the quality of performance will be limited only by the magnetic tape itself, provided, of course, that the tape deck used is of satisfactory performance. The level of total harmonic distortion in the recording process should not be greater than 0.5% with a recording current of $150\mu A$ through the head.

Equalisation to correct for head and tape characteristics is provided for each of the tape speeds: $3\frac{3}{4}$, $7\frac{1}{2}$, and 15 inches per second. The high-frequency equalisation is applied during the recording process and the low-frequency correction during playback.

Treble equalisation is achieved by using a wound Ferroxcube pot-core inductor, Mullard type WF816, in a resonant circuit between the first and second stages of the amplifier. The frequency at which maximum treble boost occurs is determined by the tuning capacitance which is adjusted by the switch SB3. The value of resistance in the feedback network to give bass equalisation during playback is selected for each tape speed by the switch SB1.

There is no provision for tone control in this pre-amplifier. It is anticipated that such control will be available with the associated amplifying equipment.

CIRCUIT DESCRIPTION

The circuit diagram of the combined record-playback pre-amplifier is given in Fig. 2. The record-playback switch SA is shown in the 'record' position and the equaliser switch SB in the position for equalisation at a tape speed of $7\frac{1}{2}$ inches per second. The playback output is taken from the anode of the second EF86, and the remaining stages are used only for the recording operation.

Controls

Two switch banks are used in the pre-amplifier. Switch SA provides the change from the recording to the playback process, and switch SB provides the

Prototype of the Tape Pre-amplifier

equalisation appropriate to one of the tape speeds: 3¾, 7½ and 15 inches per second.

The gain control RV13 is the only other control in the pre-amplifier.

Valve Complement

The complete pre-amplifier uses five Mullard valves and one Mullard germanium diode. These are:

(a) Type EF86, low noise pentode, used in the input stage.

(b) Type EF86, used in the second stage.

(c) Type EF86, used in the output stage for recording.

(d) Type ECC82, double triode, used as a push-pull oscillator for the bias and erase signals.

(e) Type EM81, tuning indicator, used in the recording-level stage.

(f) Type OA81, germanium diode, used as the indicator-circuit rectifier.

Input Stage

The pentode, type EF86, acts as a voltage amplifier for both recording and playback processes. It is possible to record from either microphone or radio sources, the radio input also being convenient for recording from crystal pick-ups. Both inputs are fed to the grid of the valve, the radio input being attenuated to the level of the microphone input. Switching is achieved by inserting the jacks, so that only one input may be used at a time.

Equaliser Stage

As was stated above, no tone control is incorporated in this stage. The output is taken during playback from across part (R10) of the anode load of the second EF86. The output supplied is 250mV at a source impedance of 15kΩ. A rearrangement of the anode load resistance (that is, R10 and R11) can be made, if required, to give an output of, for instance, 1V at a source impedance of 60kΩ.

The output of the second stage of the amplifier is fed during the recording process from the anode of the EF86 by way of the gain control RV13 to the grid of the following EF86.

A resonant circuit containing a wound Ferroxcube pot-core inductor L3 (Type WF816) is used to provide treble equalisation. The value of tuning capacitance in the resonant circuit is selected by the switch SB3 to give the maximum treble boost at frequencies appropriate to the tape speed used. The extent of treble boost is controlled by the resistor R30 and the damping resistors R34 and R35 connected in parallel with the capacitors C26 and C25. The steep rise in boost which occurs below the resonant frequency is modified by damping the inductance, and by partially shunting the resistor R30 at the appropriate frequency.

The values of the resistors R31, R32 and R33 arranged on switch SB1 have been chosen to give appropriate feedback for bass equalisation during playback. To avoid capacitive coupling, this section of the equaliser switch is arranged on the front of the switch wafer.

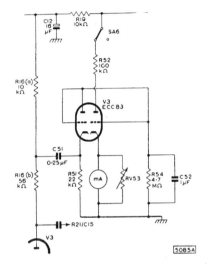

Fig. 1—Circuit for metering the recording level

Recording Output Stage

The third stage of the unit, operative only during the recording process, uses another EF86, the grid of which is fed from the volume control RV13. The stage is designed to give low harmonic distortion at peak levels of recording current, and the distortion should not exceed 0·5% for a recording current of 150μA.

The recording current is fed to the recording head by way of a parallel-T network, which acts primarily as a bias-voltage rejector circuit. The series resistance of the network is needed in this stage to ensure a constant drive to the head, and its inclusion is also desirable to preserve a satisfactory a.c./d.c. load ratio for the EF86 of the third stage.

H.F. Oscillator Stage

The halves of the double triode are arranged as a push-pull oscillator. The stage is designed so that the ECC82 draws approximately the same h.t. current during playback, when it is inoperative, as it does under its oscillatory conditions during recording. In

TAPE PRE-AMPLIFIER

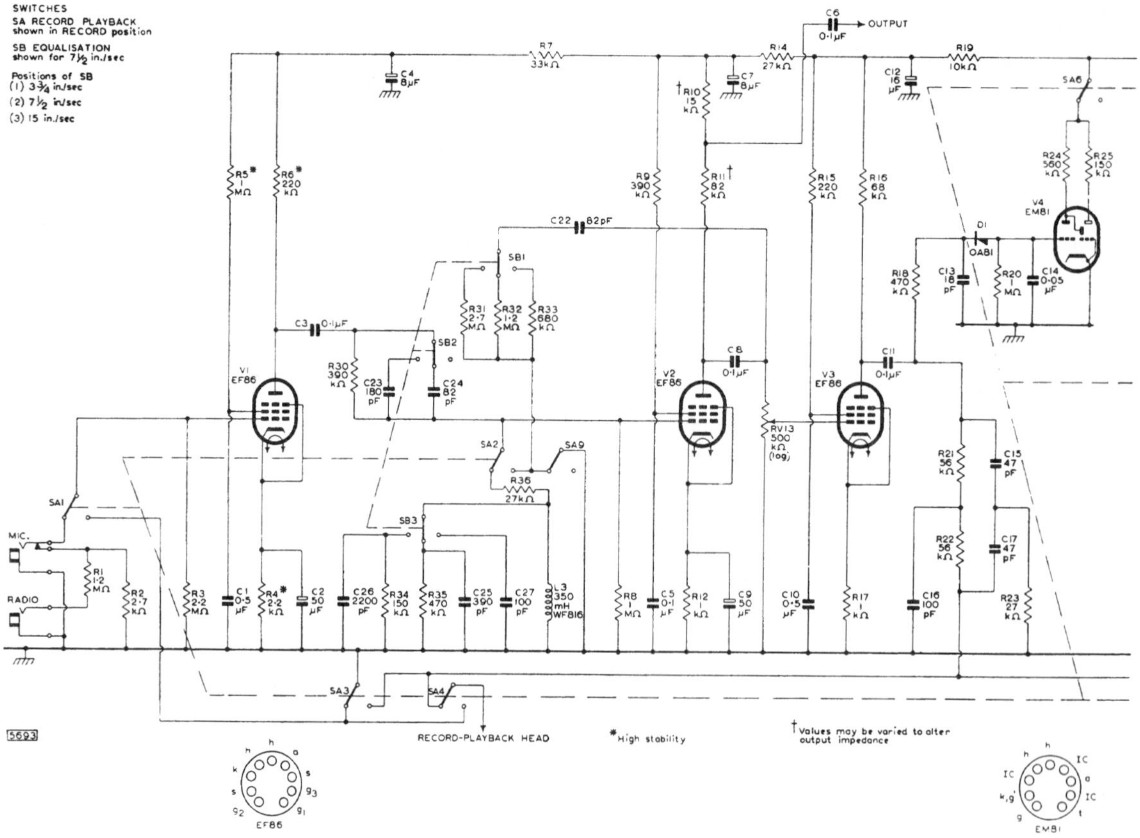

LIST OF

Resistors

Circuit ref.	Value	Tolerance (±%)	Rating (W)	Circuit ref.	Value	Tolerance (±%)	Rating (W)
R1	1·2MΩ	10	¼	R20	1 MΩ	10	¼
R2	2·7kΩ	10	¼	R21	56 kΩ	10	¼
R3	2·2MΩ	10	¼	R22	56 kΩ	10	¼
[1]R4	2·2kΩ	5	½	R23	27 kΩ	10	¼
[1]R5	1 MΩ	5	½	R24	560 kΩ	10	¼
[1]R6	220 kΩ	5	½	R25	150 kΩ	10	¼
R7	33 kΩ	10	¼	R26	4·7kΩ	10	¼
R8	1 MΩ	10	¼	R27	22 kΩ	10	¼
R9	390 kΩ	10	¼	R28	4·7kΩ	10	¼
[2]R10	15 kΩ	10	½	R29	22 kΩ	10	¼
[2]R11	82 kΩ	10	¼	R30	390 kΩ	10	¼
R12	1 kΩ	10	¼	R31	2·7MΩ	10	¼
RV13	500 kΩ logarithmic potentiometer			R32	1·2MΩ	10	¼
R14	27 kΩ	10	¼	R33	680 kΩ	10	¼
R15	220 kΩ	10	¼	R34	150 kΩ	10	¼
R16	68 kΩ	10	¼	R35	470 kΩ	10	¼
R17	1 kΩ	10	¼	R36	27 kΩ	10	¼
R18	470 kΩ	10	¼	R37	680 Ω	10	1
R19	10 kΩ	10	¼				

[1]High stability, cracked carbon

[2]Values may be adjusted to vary output impedance

TAPE PRE-AMPLIFIER

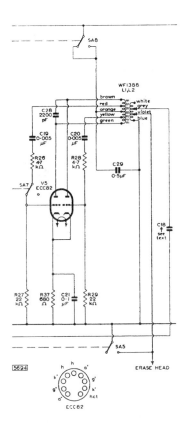

Fig. 2—Circuit diagram of tape pre-amplifier

Capacitors

Circuit ref.	Value		Description	Rating (V)
C1	0·5	µF	paper	350
C2	50	µF	electrolytic	12
C3	0·1	µF	paper	350
C4	8	µF	electrolytic	350
C5	0·1	µF	paper	350
C6	0·1	µF	paper	350
C7	8	µF	electrolytic	350
C8	0·1	µF	paper	350
C9	50	µF	electrolytic	12
C10	0·5	µF	paper	350
C11	0·1	µF	paper	350
C12	16	µF	electrolytic	350
C13	18	pF	silvered mica	
C14	0·05	µF	paper	150
C15	47	pF	silvered mica	
C16	100	pF	silvered mica	
C17	47	pF	silvered mica	
C18	*		silvered mica	
C19	0·005	µF	paper	350
C20	0·005	µF	paper	350
C21	0·1	µF	paper	350
C22	82	pF	silvered mica	
C23	180	pF	silvered mica	
C24	82	pF	silvered mica	
C25	390	pF	silvered mica	
C26	2200	pF	silvered mica	
C27	100	pF	silvered mica	
C28	2200	pF	silvered mica	
C29	0·5	µF	paper	350

Tolerance of all silvered mica capacitors is ±10%

*C18: 56pF for Brenell or Collaro decks
     47pF for Lane deck
     120pF for Motek deck
     180pF for Truvox deck

COMPONENTS

Miscellaneous

Output socket, recessed coaxial. Belling Lee, L.734/S
Record-playback head coaxial socket. Belling Lee, L.604/S
Erase head coaxial socket. Belling Lee, L.604/S
Supply input socket. Elcom, P04
Input jack (radio). Igranic, P71
Input jack (microphone). Igranic, P72
Record-playback switch SA:
    Shirley Laboratories Ltd., 16370/B3
    Specialist Switches, SS/567/A
(Note: Details of proprietary switches may not be identical with those given in the diagrams.)
Equaliser switch SB, 3-pole, 3-way
Five-way tagboard (one). Bulgin, C120. Denco
Ten-way tagboard (three). Bulgin, C125. Denco.
Ceramic stand-off insulator (two)
Stand-off insulator, 2-tag

Inductors

Equalisation coil: Wound Ferroxcube pot-core inductor. Mullard, WF816
Oscillator coil: Wound Ferroxcube pot-core inductor. Mullard, WF1388

Valves and Germanium Diode

V1	Low noise pentode,	Mullard type EF86
V2	Low noise pentode,	Mullard type EF86
V3	Low noise pentode,	Mullard type EF86
V4	Tuning indicator,	Mullard type EM81
V5	Double triode,	Mullard type ECC82
D1	Germanium diode,	Mullard type OA81

Valveholders

B9A valveholder (two). McMurdo, BM9/U
B9A nylon-loaded valveholder with screening skirt (two). McMurdo, XM9/AU; skirt 95
B9A nylon-loaded valveholder with screening skirt and flexible mounting. McMurdo, XM9/UXG.1; skirt 95

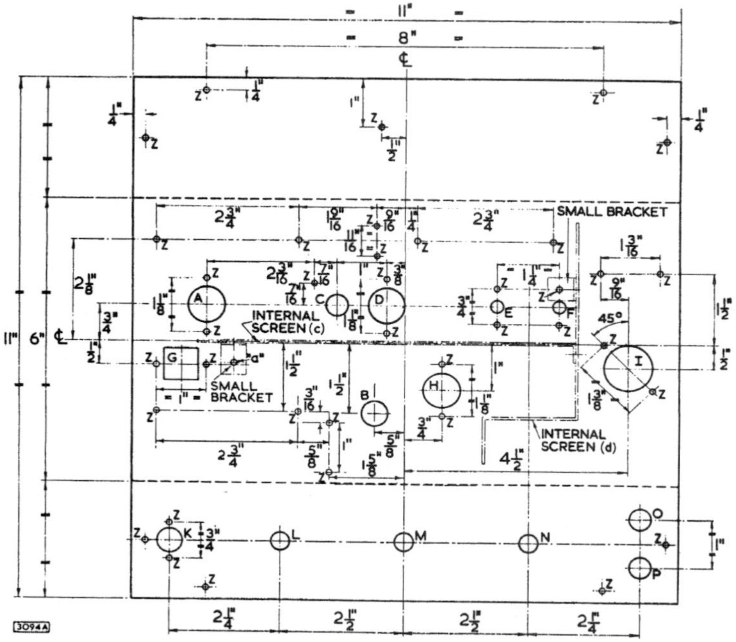

(a) (above) Main chassis (hole marked 'a' should be made after assembly with No. 34 drill)

(b) (below) Base

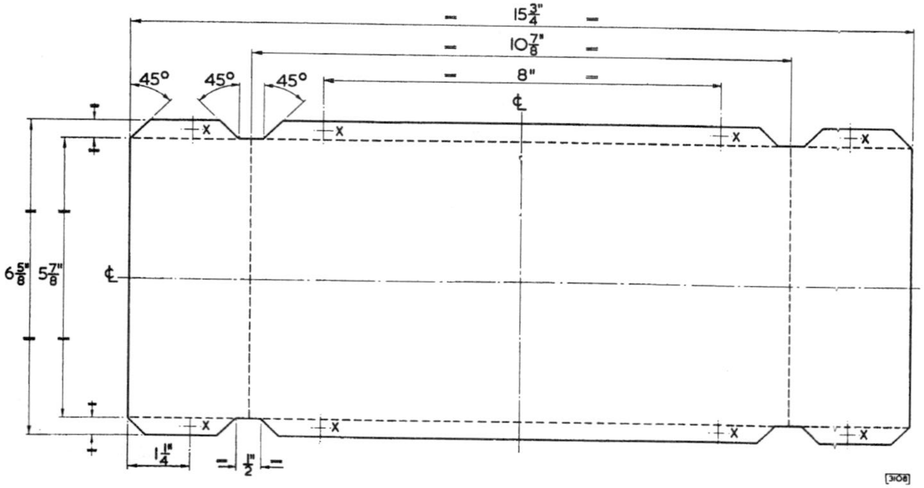

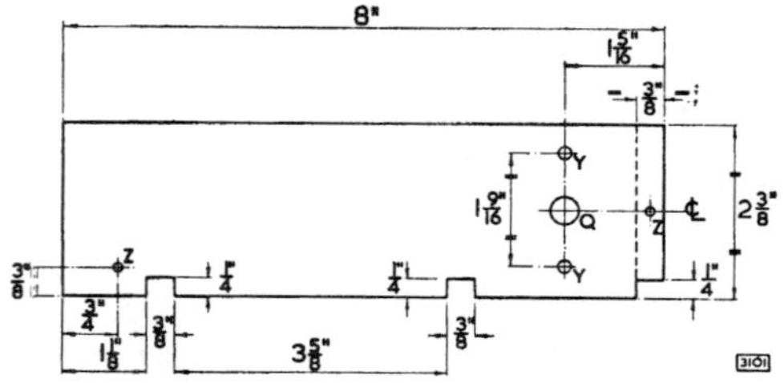

(c) (above) Internal screen

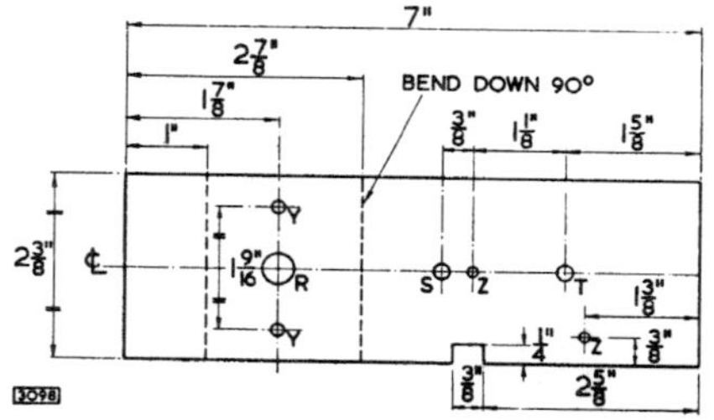

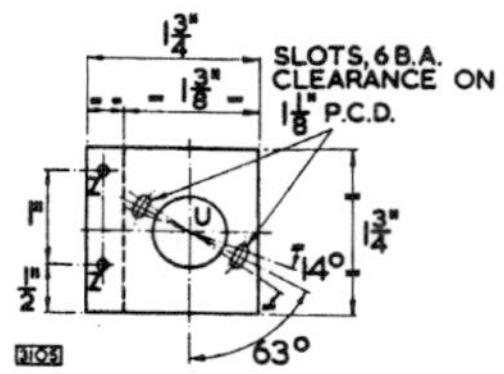

(e) (above) EM81 fixing bracket

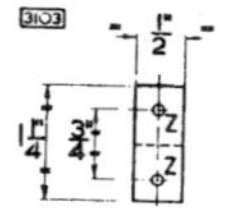

(f) (above) Small bracket

(d) (left) Internal screen

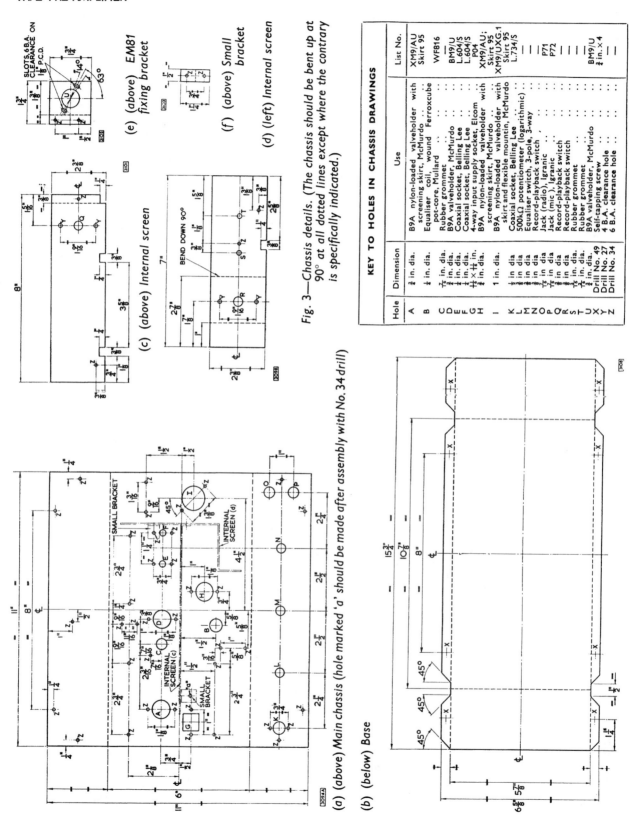

Fig. 3—Chassis details. (The chassis should be bent up at 90° at all dotted lines except where the contrary is specifically indicated.)

KEY TO HOLES IN CHASSIS DRAWINGS

Hole	Dimension	Use	List No.
A	$\frac{3}{4}$ in. dia.	B9A nylon-loaded valveholder with screening skirt, McMurdo	XM9/AU Skirt 95
B	$\frac{1}{4}$ in. dia.	Equaliser coil, wound Ferroxcube pot-core, Mullard	WF816
C	$\frac{7}{16}$ in. dia.	Rubber grommet	—
D	$\frac{3}{4}$ in. dia.	B9A valveholder, McMurdo	BM9/U
E	$\frac{1}{4}$ in. dia.	Coaxial socket, Belling Lee	L.604/S
F	$\frac{1}{4}$ in. dia.	Coaxial socket, Belling Lee	L.604/S
G	$\frac{11}{16} \times \frac{11}{16}$ in.	4-way input supply socket, Elcom ..	P04
H	$\frac{3}{4}$ in. dia.	B9A nylon-loaded valveholder with screening skirt, McMurdo	XM9/AU; Skirt 95
I	1 in. dia.	B9A nylon-loaded valveholder with skirt and flexible mountin, McMurdo	XM9/UXG.1 Skirt 95
K	$\frac{1}{2}$ in dia	Coaxial socket, Belling Lee	L.734/S
L	$\frac{3}{8}$ in dia	500kΩ potentiometer (logarithmic) ..	—
M	$\frac{3}{8}$ in dia	Equaliser switch, 3-pole, 3-way ..	—
N	$\frac{3}{8}$ in dia	Record-playback switch	—
O	$\frac{7}{16}$ in dia	Jack (radio), Igranic	P71
P	$\frac{7}{16}$ in dia	Jack (mic), Igranic	P72
Q	$\frac{3}{8}$ in dia	Record-playback switch	—
R	$\frac{3}{8}$ in dia	Record-playback switch	—
S	$\frac{3}{16}$ in. dia.	Rubber grommet	—
T	$\frac{3}{16}$ in. dia.	Rubber grommet	—
U	$\frac{3}{4}$ in. dia.	B9A valveholder, McMurdo	BM9/U
X	Drill No. 49	Self-tapping screw	$\frac{3}{8}$ in. × 4
Y	Drill No. 27	4 B.A. clearance hole	—
Z	Drill No. 34	6 B.A. clearance hole	—

TAPE PRE-AMPLIFIER

this way, the total current drain for either the recording or the playback process does not alter greatly, and the design of the power supply is consequently simplified.

The oscillator coil consists of a wound Ferroxcube pot-core, type WF1388, in which the secondary windings are tapped so that the coil can be used with most of the commercially available tape decks. The value of the coupling capacitor C18 will vary with the type of record-playback head used, and a list of appropriate values is given in Table 1.

Power Supply
The unit described in Chapter 11 is suitable for use with the pre-amplifier circuit. The h.t. and heater requirements of the two tape circuits are similar.

CONSTRUCTION AND ASSEMBLY
The chassis for the tape pre-amplifier is made up of seven separate pieces of 16 s.w.g. aluminium sheet. The dimensions (in inches) of these are:

(a) Main chassis 11×11
(b) Base $15\frac{3}{4} \times 6\frac{5}{8}$
(c) Internal screen $8 \times 2\frac{3}{8}$
(d) Internal screen $7 \times 2\frac{3}{8}$
(e) EM81 mounting bracket $1\frac{3}{4} \times 1\frac{1}{4}$
(f) Small bracket (two) $1\frac{1}{4} \times \frac{1}{2}$

Each piece should be marked as shown in the chassis drawings of Fig. 3, and the holes should be cut as indicated. It is important that, when bending the sheet, the scribed lines should be exactly along the angles. This ensures that the pieces will fit together properly when assembled.

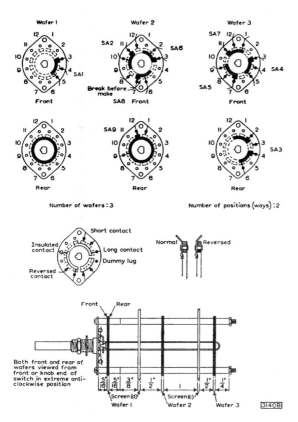

Fig. 4—Record-playback switch details

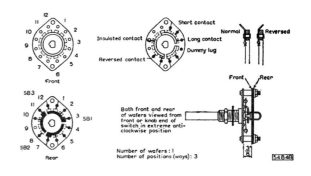

Fig. 5—Equaliser switch details

Recording Level Indicator
The tuning indicator, type EM81, used in this stage is fed from the anode of the recording stage. The large series resistance R18 is used to minimise the loading of the recording stage. If a meter indication of the recording level is required, the circuit given in Fig. 1 can be used in place of the EM81. This alternative arrangement is similar to that suggested on page 92 for the 3W tape amplifier.

Details of the record-playback switch SA are given in Fig. 4. Before assembling this switch around the screens, it will probably be found convenient to fix the following components to the chassis:

(1) The erase and record-playback coaxial sockets which have to be fitted to the chassis beneath wafer 3 of the switch.

(2) All the valveholders. Only the three EF86 valves should be skirted, and the holders for these valves should be nylon-loaded. The holder for the input valve V1 (EF86) should be of an anti-microphonic type – that is, having a flexible mounting.

(3) The two small brackets f, which should be bolted

TAPE PRE-AMPLIFIER

to the internal screens c and d as shown in the layout diagram in Fig. 10.

The construction is best continued after the above components have been fitted by assembling wafers 1 and 2 around the internal screens. The wafers should be arranged so that positions 6 and 7 are nearest the chassis and the face of each wafer described as 'rear' in the switch diagram (Fig. 4) is farthest from the switch plate. The internal screen c should be added to the assembly, both screens should be bolted together, and wafer 3 should be fitted in position, again with its 'rear' face farthest from the switch plate. The general arrangement of the switch wafers and internal screens is shown in Fig. 10. The use of a shake-proof washer is recommended between the switch plate and the front panel when the switch and screens are fitted to the main chassis. The arrangement of the smaller components on tagboards is shown in Figs. 6 to 9, and a suitable layout of these boards and the other components in the chassis is shown in Fig. 10. In fitting the EM81 mounting bracket, the valveholder should be fastened to the bracket so that the solder tags are on the same side as the flange on the bracket. The gap between pins 1 and 9 on the holder should face the flange to ensure that the tuning eye of the EM81 faces forward. The indicator should appear in the centre of the front panel of the equipment, and the bracket should be bolted to the chassis in the correct position for this to be so. The valveholder can be fixed in the main chassis, as described in Chapter 11, if it is required to mount the EM81 at the end of extension leads. An extra hole with a diameter of $\frac{3}{4}$ in. will have to be drilled for the valveholder, in place of the grommet hole C shown in Fig. 3(a).

Identification of the coil winding suitable for the erase heads of various decks can be made by referring to Table 1. The appropriate value of C18 is also given in this table. The bias capacitor is connected to the *grey* lead from the secondary winding for all the makes of deck indicated. The connections in the diagram for the primary winding of the coil and for the blue (earth) lead from the secondary winding are also the same for the decks listed.

TABLE 1
Circuit Arrangement for Various Tape Decks

Tape Deck	Erase Lead	C18 (pF)
Brennel	violet	56
Collaro	violet	56
Lane	grey	47
Motek	grey	120
Truvox	white	180

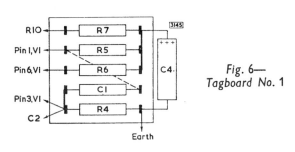

Fig. 6—Tagboard No. 1

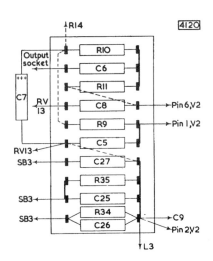

Fig. 7—Tagboard No. 2

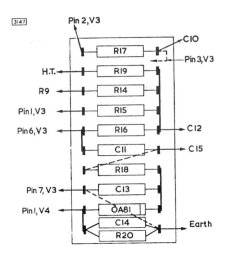

Fig. 8—Tagboard No. 3

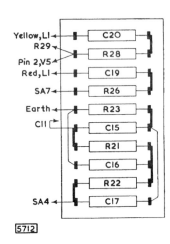

Fig. 9—Tagboard No. 4

TAPE PRE-AMPLIFIER

Underside View of Prototype Pre-amplifier

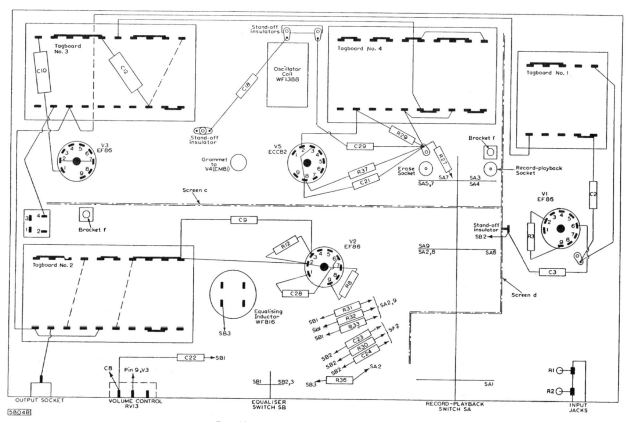

Fig. 10—Suggested layout of components

TAPE PRE-AMPLIFIER

PERFORMANCE

Frequency Response

The overall response of the recorder depends on the type of head used, the magnitude of the bias current, and, to some extent, on the tape employed. The low-frequency response depends on the amplifier used rather than on the type of head, and in this design, it will not be more than 3dB down at 50c/s. The high-frequency response depends on the tape speed and the gap width of the head. With heads having a gap width of 0·005 in., the following performance figures (relative to the level at 1kc/s) can be obtained:

15 in./sec   ±3dB from 50c/s to 15kc/s
7½ in./sec   ±3dB from 50c/s to 12kc/s
3¾ in./sec   ±3dB from 50c/s to 5kc/s

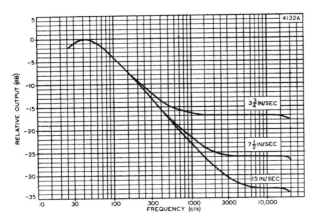

Fig. 11—Bass-boost characteristics

Flexibility has been achieved in the treble boost circuits of the pre-amplifier. Consequently good control of the complete response of the pre-amplifier is possible and equalisation to reasonably high frequencies is practicable. However, the degrees of control and equalisation depend on the individual adjustments of component values to suit the head and tape being used. The component values given in this chapter apply only to the pre-amplifier when it is used with E.M.I. recording tape and a Collaro tape deck.

Ferroxcube pot-core inductors are adequately screened to prevent excessive hum or stray bias being picked up, and they do not appear to cause more circuit-ringing than RC networks which produce the same treble boost.

The playback characteristic of the pre-amplifier conforms to the C.C.I.R. specification, thus permitting excellent reproduction of pre-recorded tapes. The recording characteristic is arranged to give a flat frequency response in conjunction with this playback characteristic. Additional head losses occurring during playback will normally be capable of correction by the tone controls in the associated amplifying systems.

Sensitivity

The sensitivity of the recording process is measured with the control RV13 set for maximum gain. This does not apply to the feedback sensitivity because the gain control is not operative at the point from which the output is taken for the associated equipment.

Recording Sensitivity
(measured at 1kc/s, with recording-head audio current of 150μA)

(a) Microphone input: 0·5mV for peak
    (impedance = 2MΩ)   recording level
(b) Radio input: 250mV for peak
    (impedance = 1·2MΩ) recording level

Playback Sensitivity
(measured at 5kc/s for each tape speed for output of 250mV)

(a) 15 in./sec: 5·5mV
(b) 7½ in./sec: 2·4mV
(c) 3¾ in./sec: 1·0mV

TEST PROCEDURE

The four tests outlined below are intended as simple, yet quite effective, checks for the pre-amplifier.

The values given in the various tables and figures were obtained from the prototype pre-amplifier, using Collaro record-playback and erase heads. The bias current used throughout was 1·0mA at a frequency of 60kc/s, and the erase-head voltage was about 25V, again at a frequency of 60kc/s.

Test I—D.C. Voltages

The d.c. voltages at points in the equipment should be tested with reference to Table 2. The results shown in this table were obtained using an Avometer No. 8.

Test II—Amplifier on Playback

Two pieces of equipment are required for this test:

(1) A signal generator covering a frequency range from 20c/s to 20kc/s;
(2) A valve voltmeter covering a frequency range from 20c/s to 20kc/s.

TABLE 2

D.C. Conditions

Point of Measurement		Voltages (V)		D.C. Range of Avometer* (V)
		(a) SA in Record position	(b) SA in Playback position	
	C12	250	250	1000
	C7	190	190	1000
	C4	163	163	1000
V5 (ECC82)	2nd Anode	290	290	1000
	1st Anode	290	290	1000
	1st Cathode	14	12	1000
V4 (EM81)	Anode	40	0	1000
	Target	145	0	1000
V3 (EF86)	Anode	110	110	1000
	Screen grid	140	140	1000
	Cathode	2·6	2·6	10
V2 (EF86)	Anode	60	60	1000
	Screen grid	90	90	1000
	Cathode	1·5	1·5	10
V1 (EF86)	Anode	50	50	1000
	Screen grid	60	60	1000
	Cathode	1·4	1·4	10

*Resistance of Avometer:
1000V-range, resistance = 20MΩ;
100V-range, resistance = 2MΩ;
10V-range, resistance = 200kΩ.

TABLE 3

Playback Sensitivity

(Signal frequency = 5kc/s)

Tape Speed (in./sec)	Input (mV)	Output (mV)
15	5·5	250
7½	2·4	250
3¾	1·0	250

The record-playback switch SA should be in the playback position. A signal from the generator, having a frequency of 5kc/s, should be applied to the record-playback socket (which normally accommodates the connecting plug from the record-playback head). The consequent output signal should be measured on the voltmeter at the output socket.

The input voltage should be adjusted to give an output voltage at the output socket of 250mV for each tape speed, and the input required for this output should be noted. The voltage readings that should be obtained are given in Table 3.

For operation at such high sensitivities, great care should be taken to ensure that the signal measured is not composed mostly of hum. It is advisable, therefore (a) to use screening cans on the three EF86s, (b) to screw on firmly the base of the amplifier, and (c) to use coaxial cables for the connections to the measuring equipment.

The input voltage at 5kc/s should be varied until the output voltage drops to 50mV. The frequency of the signal should then be reduced to 40c/s and the values of boost listed in Table 4 should be observed at the output socket.

The bass-boost characteristics for the three tape speeds are shown in Fig. 11.

Test III—Amplifier on Record

The instruments required for this test are:

(1) A signal generator covering a frequency range from 20c/s to 20kc/s;

(2) A valve voltmeter[1] covering a frequency range from 20c/s to 20kc/s.

The record-playback and erase heads should be connected to the appropriate sockets in the pre-amplifier, and the equipment should be switched to the recording condition.

A signal at 1kc/s should be applied from the generator to the radio input socket. The magnitude of this signal should be such that an output of 15mV is obtained at the output socket.

The boost indicated in Table 5 should be obtained at the appropriate tape speed when the signal frequency is altered to the value shown in the table.

---

[1] For accurate results, two separate pieces of p.v.c. covered wire are recommended for the connections to the valve voltmeter. A coaxial cable may result in considerable errors in the measurements because of the parallel capacitance which is introduced.

TAPE PRE-AMPLIFIER

The treble-boost characteristics for the three tape speeds are shown in Fig. 12.

Values for the recording sensitivity for an output voltage measured at the anode of V3 (EF86) are given in Table 6. A test of the recording-level indicator should show that the EM81 'closes' for each tape speed with approximately 15V at the anode of the recording-output valve.

An alternative method of checking the recording amplifier is possible: for each tape speed, the voltage developed across a 50Ω resistor connected in series with the recording head can be observed for the full range of signal frequencies. The response figures so obtained should agree with the values obtained with the prototype pre-amplifier and listed on page 110.

TABLE 4

Bass Boost

Signal frequency = 40c/s

(Output voltage for 5kc/s = 50mV)

Tape speed (in./sec)	Voltmeter reading (V)	Output boost (dB)
15	2·3	33
7½	1·0	26
3¾	0·36	17

TABLE 5

Treble Boost

(Output voltage for 1kc/s = 15mV)

Tape speed (in./sec)	Signal frequency (kc/s)	Voltmeter reading (mV)	Output boost (dB)
15	17	84	15
7½	12	150	20
3¾	5·5	100	16·8

For these observations, it will be necessary to disconnect one end of the resistor R26, otherwise only the bias signal will be measured.

Test IV—Bias Level

For this test, two pieces of equipment are required:

(1) A valve voltmeter which will indicate accurately at frequencies of up to 70kc/s;

(2) A resistor of 50Ω.

The resistor should be soldered in series with the earthy end of the record-playback head, and the voltage developed across the resistor, with no input signal, should be measured with the voltmeter.

The voltage developed across the resistor should be 50mV, which corresponds to a bias current of 1·0mA flowing in the 50Ω-resistor.

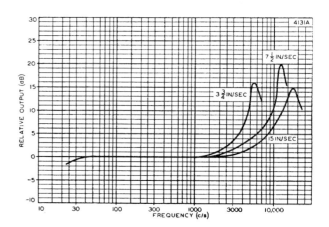

Fig. 12—Treble-boost characteristics

TABLE 6

Recording Sensitivity

Signal frequency	1	kc/s
Tape speed	15, 7½ or 3¾	in./sec
Voltage at anode of V3	15	V
Microphone input	0·5	mV
Radio input	65	mV

CHAPTER 13
Seven-watt Stereophonic Amplifier

The circuit diagram drawn in Fig. 1 is for a high-quality, dual-channel amplifier designed principally for stereophonic reproduction but also providing facilities for monaural applications. Only one channel of the amplifier is drawn. The circuitry appearing between the dotted lines is for the left-hand channel: it should be duplicated for the right-hand channel. The circuitry drawn outside the dotted lines (the power supply, for example) is common to both channels.

The total complement of valves used in the amplifier consists of one double triode, type ECC83, four triode pentodes, type ECL82, and one full-wave rectifier, type EZ81. The double triode is shared between the two channels, one section of the valve being used in each channel for voltage amplification. Two of the ECL82s are used in each channel. The triode sections of these valves form a phase-splitting stage and the pentode sections are arranged as a push-pull output stage with distributed loading. The EZ81 forms a conventional power supply with resistance-capacitance smoothing, providing the h.t. for both channels.

The rated output-power reserve of each channel is 7W, at which level the total harmonic distortion is always better than 0·5%. The low level of distortion is achieved by using 21dB of negative feedback, the feedback voltage being taken from the secondary winding of the output transformer in each channel to the cathode circuit of the corresponding input stage. The sensitivity of the circuit, even with this high value of feedback, is 100mV, which is ample for use with existing stereophonic crystal pick-up heads.

CIRCUIT DESCRIPTION
Resistors and capacitors appearing in the left-hand channel of the amplifier are numbered 1, 2, 3, etc.; the corresponding components in the right-hand channel are numbered 101, 102, 103. etc.

Prototype of Seven-watt Stereophonic Amplifier

7W STEREOPHONIC AMPLIFIER

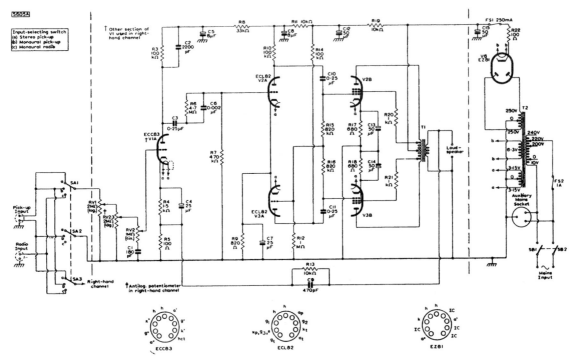

Fig. 1—Circuit diagram of seven-watt stereophonic amplifier (left-hand channel only is shown: circuitry between vertical dotted lines should be repeated in right-hand channel)

Input Selector Switch

The input stages of both channels are connected to the 3-way selector switch SA. The switch positions indicated in Fig. 1 provide the following facilities:

(a) Stereophonic reproduction from stereophonic crystal pick-up heads.

(b) Dual-channel monaural reproduction from a monaural pick-up head. In this position, the left-hand pick-up input terminal is 'live', and both channels are connected in parallel at the input. The input terminal for the right-hand channel of a stereophonic pick-up head is earthed at position b of SA2. If position b of SA3 is earthed instead of being connected to position b of SA1, single-channel reproduction is possible.

(c) Dual-channel monaural reproduction from an f.m. tuner unit. The input socket in Fig. 1 is connected for monaural applications. If position c of SA3 is connected to the right-hand input terminal instead of the left-hand terminal, the circuit will be suitable for reproducing stereophonic transmissions. If position c of SA3 is earthed instead of being connected to the input socket, the system gives single-channel monaural reproduction from an f.m. tuner unit.

Input Stage

The triode sections of the ECC83 are used for voltage amplification, one section being used in each channel. A simple treble tone-control network consisting of the components RV2 and C1 (RV102 and C101) is incorporated in the grid circuit giving continuously variable treble cut. A volume control RV1 (RV101) is also included in the grid circuit. Dual-ganged potentiometers are used in the prototype for the tone-control components RV2 and RV102 and also for the volume-control components RV1 and RV101, so that equal adjustments can be made simultaneously to both channels. It may be found more convenient to use dual-concentric potentiometers for these controls because adjustments can be made separately to each channel with them.

Instability can result from the large amount of negative feedback introduced to the cathode circuit of each input stage, and the simplest method of ensuring stability is to reduce the loop gain of each channel. To achieve this reduction at high audio frequencies, the capacitor C2 (C102) is connected to bypass the anode load resistor R3 (R103). To reduce the loop gain at bass frequencies, the parallel combination R6 and C6 (R106 and C106) is included in

7W STEREOPHONIC AMPLIFIER

LIST OF COMPONENTS

Resistors and capacitors in the left-hand channel are numbered 1, 2, 3, etc.; corresponding components in the right-hand channel are numbered 101, 102, 103, etc.

Resistors

Circuit ref.		Value	Tolerance (±%)	Rating (W)
*RV1 and RV101		2×2 MΩ logarithmic potentiometer		
*RV2 and RV102		2×1 MΩ linear potentiometer		
R3,	R103	100 kΩ	10	½
R4,	R104	1·5 kΩ	10	½
R5,	R105	100 Ω	10	¼
R6,	R106	4·7 MΩ	10	¼
R7,	R107	470 kΩ	10	¼
R8,	R108	33 kΩ	10	½
R9,	R109	820 Ω	10	½
R10,	R110	100 kΩ	10	¼
R11,	R111	10 kΩ	10	½
R12,	R112	1 MΩ	10	¼
R13,	R113	10 kΩ	10	¼
R14,	R114	100 kΩ	10	¼
R15,	R115	820 kΩ	5	¼
R16,	R116	820 kΩ	5	¼
R17,	R117	680 Ω	10	2
R18,	R118	680 Ω	10	2
R19,	R119	10 kΩ	10	½
R20,	R120	1 kΩ	10	½
R21,	R121	1 kΩ	10	½
	R22	100 Ω	10	6
*RV23		2 MΩ logarithmic potentiometer		
*RV123		2 MΩ antilogarithmic potentiometer		

*10% law (See text)

Capacitors

Circuit ref.		Value		Description	Rating (V)
C1,	C101	180	pF	silvered mica	
C2,	C102	2200	pF	silvered mica	
C3,	C103	0·25	μF	paper	275
C4,	C104	25	μF	electrolytic	3
C5,	C105	8	μF	electrolytic	300
C6,	C106	0·002	μF	paper	275
C7,	C107	25	μF	electrolytic	3
C8,	C108	8	μF	electrolytic	300
C9,	C109	470	pF	silvered mica	
C10,	C110	0·25	μF	paper	275
C11,	C111	0·25	μF	paper	275
†C12,	†C112	50	μF	electrolytic	300
C13,	C113	50	μF	electrolytic	3
C14,	C114	50	μF	electrolytic	3
†C15		50	μF	electrolytic	300

†(50+50+50)μF electrolytic capacitor
Tolerance of silvered mica capacitors is ±10%

Mains Transformer

Primary: 10–0–200–220–240V
Secondaries: H.T. 250–0–250V, 150mA
L.T. 3·15–0–3·15V, 4A (for ECC83 and ECL82s)
0–6·3V, 1A (for EZ81)

Output Transformer

Primary: 9kΩ
Secondary 3·75, 7·5 and 15Ω

Commercial Components

Manufacturer	Type No.
Colne	03062
Elden	1298
Gardners	AS.7009
Gilson	W.O.1340
Hinchley	1535
Howells	MO10
Parmeko	P2932
Partridge	P4133
Wynall	7914

Commercial Components

Manufacturer	Type No.
Colne	04022
Elden	1297
Gardners	RS.3110
Gilson	W.O.741B
	W.O.741A/B
Hinchley	1447
Howells	MM10
Parmeko	P2931
Partridge	P4132
Savage	5D32
Wynall	7915

Valves

Mullard ECC83, ECL82 (four), EZ81

Valveholders

B9A (noval) nylon-loaded with screening skirt (for ECC83)
McMurdo XM9/AU, skirt 95
B9A (noval) (five), McMurdo BM9/U

Miscellaneous

Mains input plug, 3-pin. Bulgin, P429
Mains switch. N.S.F., 8370/B3
Fused voltage selector. Clix, VSP/393/2
H.T. supply plug (pre-amplifier) 6-pin. Bulgin, P427
Fuseholder, Minifuse. Belling Lee, L.575
Fuse, 250mA
Lampholder. Bulgin, D675/1/Red
Pilot lamp. 6·3V, 0·15A., L.E.S.
Input socket, 3-pin. Belling Lee, L.790/CS
Output plug, (two) 2-pin, non-reversible. Painton, 313263

the coupling circuit between the first and second stages of each channel. This manner of coupling results in resistive attenuation at low audio frequencies, so that the phase shift is limited.

Balance Control

To compensate for any differences in acoustical output resulting from unequal outputs from the stereophonic pick-up head or unequal sensitivities of the loudspeakers, a balance control consisting of RV23 and RV123 is incorporated between the volume and tone controls. This is made up of a logarithmic potentiometer connected in reverse in one channel and an antilogarithmic potentiometer connected normally in the other. The characteristics of this control are discussed on page 28 in Chapter 4. If dual-concentric potentiometers are used for the volume control RV1, RV101, balance can be achieved with them, and the potentiometers RV23, RV123 can be omitted.

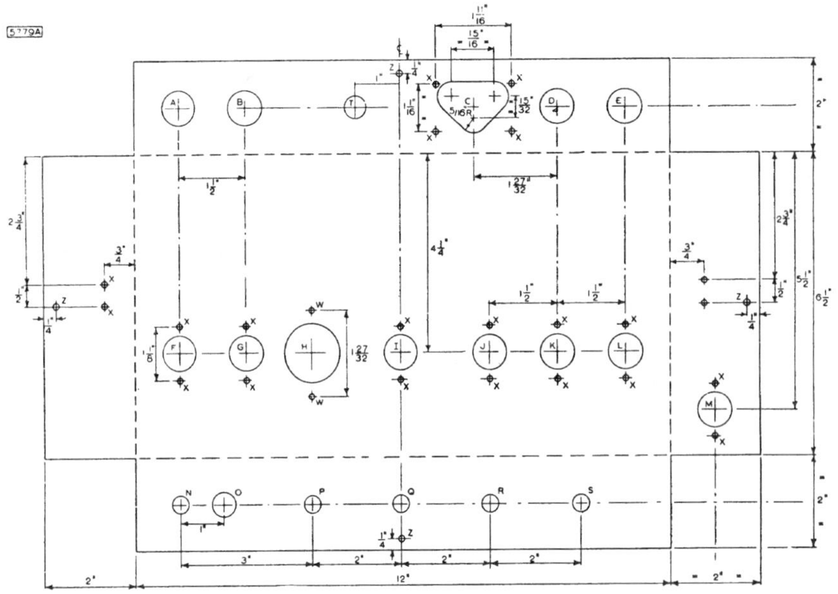

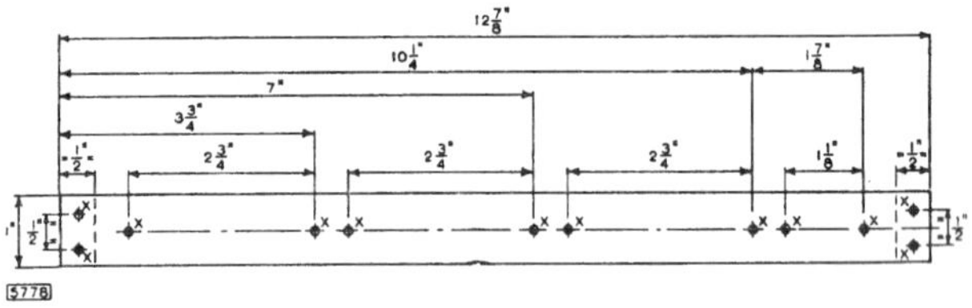

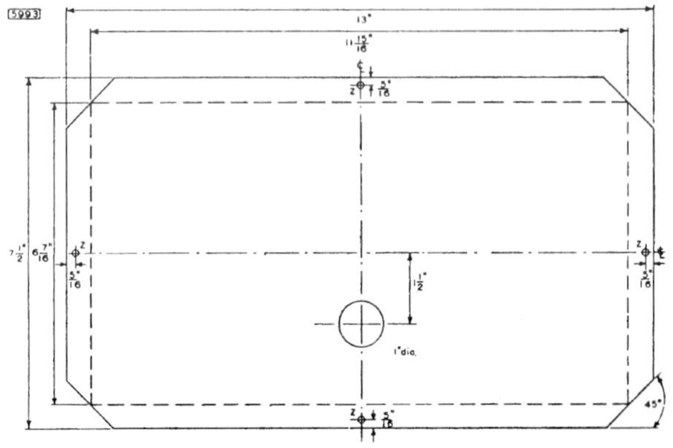

Fig. 2—Chassis details (the pieces should be bent up at 90° at all dotted lines)
(a) (above) Main chassis
(b) (right) Tagboard mounting strip
(c) (bottom right) Cover plate

KEY TO HOLES IN CHASSIS

Hole	Dimension	Use	Type No.
A	$\frac{3}{4}$ in. dia.	Output socket, 2-pin. Painton	313263
B	$\frac{3}{4}$ in. dia.	Mains input plug. 3-pin. Bulgin ..	P429
C	—	Fused voltage selector. Clix	VSP/393/2
D	$\frac{3}{4}$ in. dia.	H.T. supply plug, 6-pin. Bulgin.. ..	P427
E	$\frac{3}{4}$ in. dia.	Output socket, 2-pin. Painton	313263
F	$\frac{3}{4}$ in. dia.	B9A (noval) valveholder. McMurdo ..	BM9/U
G	$\frac{3}{4}$ in. dia.	B9A (noval) valveholder. McMurdo ..	BM9/U
H	$1\frac{1}{4}$ in. dia.	Electrolytic capacitor	—
I	$\frac{3}{4}$ in. dia.	B9A (noval) nylon-loaded valveholder with screening skirt. McMurdo ..	XM9/AU Skirt 95
J	$\frac{3}{4}$ in. dia.	B9A (noval) valveholder. McMurdo ..	BM9/U
K	$\frac{3}{4}$ in. dia.	B9A (noval) valveholder. McMurdo ..	BM9/U
L	$\frac{3}{4}$ in. dia.	B9A (noval) valveholder. McMurdo ..	BM9/U
M	$\frac{3}{4}$ in. dia.	Input socket, 3-pin. Belling Lee ..	L.790/CS
N	$\frac{3}{4}$ in. dia.	Lampholder. Bulgin	D675/1/Red
O	$\frac{1}{2}$ in. dia.	Mains switch. N.S.F.	8370/B3
P	$\frac{3}{8}$ in. dia.	2×1MΩ logarithmic potentiometers ..	—
Q	$\frac{3}{8}$ in. dia.	500kΩ log/antilog potentiometers ..	—
R	$\frac{3}{8}$ in. dia.	2×100kΩ logarithmic potentiometers	—
S	$\frac{3}{8}$ in. dia.	Selector switch, 3-way, miniature rotary	—
T	$\frac{7}{16}$ in. dia.	Fuseholder, Minifuse. Belling Lee ..	L.575
X	2·9 mm dia.	—	—
Z	Drill No. 34	6 B.A. clearance holes	—

7W STEREOPHONIC AMPLIFIER

Phase-splitting Stage
The output from the anode of V2A is taken to the grid of V3A by way of the resistors R15 and R12 (R115 and R112). These components also form the grid-leak resistance for the pentode section V2B, and R16 and R12 (R116 and R112) comprise the grid-leak resistance for V3B. Balance between the output voltages from the triodes of the phase-splitter is governed by the values of these three resistors.

Output Stage
The pentode sections of the ECL82s are used in a push-pull output stage. Distributed loading is used, and the primary winding of the output transformer is tapped so that 20% of the winding in each anode circuit also appears in the corresponding screen-grid circuit. (The choice of tapping points is discussed in Chapter 3.)

Negative Feedback
The amount of negative feedback which is used between the secondary winding of the output transformer in each channel to the cathode circuit of the input stage of that channel is 21dB. The output resistance of each channel with this feedback is 0·54Ω measured at the 15Ω output terminal. This gives an adequate damping factor of approximately 28.

Power Supply
A conventional power supply using the Mullard full-wave rectifier, type EZ81, with resistance-capacitance smoothing, provides the h.t. for both channels. Both supply connections are made to the common reservoir capacitor C15, but smoothing is achieved separately in each channel with the components R19, C12 and R119, C112.

The resistor R22 in the cathode circuit of the EZ81 should, with the transformer resistance, provide the minimum limiting resistance quoted for the valve. The choice of value for limiting resistors is discussed on page 28. A total h.t. current of 150mA at 260V is required for the amplifier, and the total heater current needed is 5A at 6·3V.

CONSTRUCTION AND ASSEMBLY

The chassis for the 7W stereophonic amplifier is made from three separate pieces of 16 s.w.g. aluminium sheet. The dimensions (in inches) of these pieces are:

(a) Main chassis 16 ×10½
(b) Tagboard mounting strip 12⅞×1
(c) Cover plate 13 ×7½

Each piece should be marked as shown in the drawings of Fig. 2, and the holes should be cut as indicated. Where bending is required, it is important that the bends should be made accurately at the lines for the pieces to fit together properly on assembly.

Most of the smaller components should be mounted on tagboards, the wiring of which is shown in Figs. 4 to 7, and the tagboards should be fixed to the mounting strip. The larger components such as switches, valveholders and transformers should be mounted in the chassis in the positions indicated in the layout diagram of Fig. 3. The recommended wiring between these components is also indicated in Fig. 3. It will be seen that the two channels of the amplifier are separate, one channel being in each half of the chassis.

PERFORMANCE

Distortion
The total harmonic distortion was measured in the prototype amplifier with a continuous sine-wave input signal at 400c/s. With 21dB of negative feedback, the distortion for the rated output of 7W per channel

TABLE 1
D.C. Conditions in Each Channel

	Point of Measurement	Voltage (V)	Range of D.C. Avometer* (V)
	C15 Common to both channels	260	1000
	C12	230	1000
	C8	210	1000
	C5	190	1000
V2, V3 ECL82	Pentode anodes	256	1000
	Pentode screen grids	255	1000
	Pentode cathodes	21	100
	Triode anode V3	130	1000
	Triode anode V2	110	1000
	Triode cathodes	2·0	25
V1 ECC83	Anode	90	100
	Cathode	1·5	25

*Resistance of Avometer:
1000V-range, resistance = 20MΩ
100V-range, resistance = 2MΩ
25V-range, resistance = 500kΩ

7W STEREOPHONIC AMPLIFIER

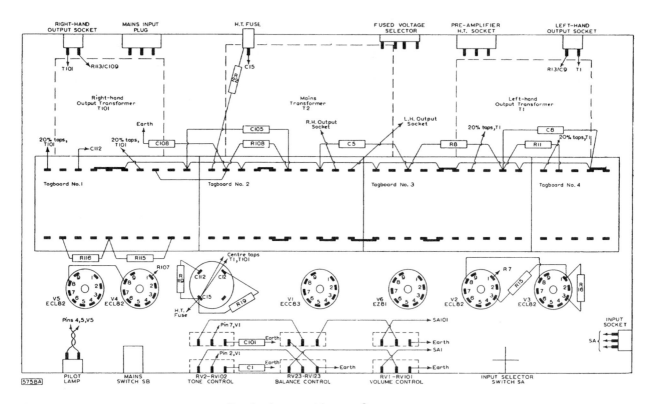

Fig. 3—Suggested layout of components

Underside View of Prototype Amplifier

7W STEREOPHONIC AMPLIFIER

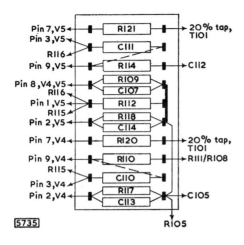

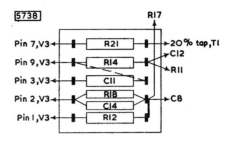

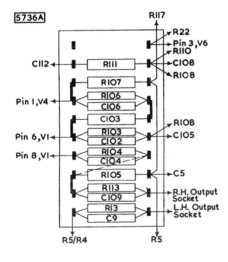

Fig. 4 (top left)—Tagboard No. 1

Fig. 5 (middle left)—Tagboard No. 2

Fig. 6 (bottom left)—Tagboard No. 3

Fig. 7 (above)—Tagboard No. 4

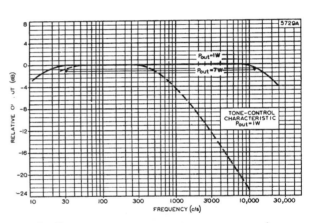

Fig. 8—Frequency-response, power-response and tone-control characteristics

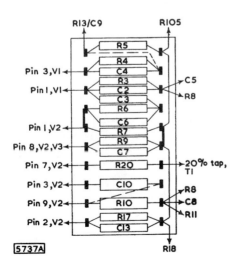

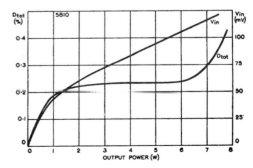

Fig. 9—Variations in harmonic distortion D_{tot}, and input V_{in} with output power

7W STEREOPHONIC AMPLIFIER

is always better than 0·5%. The variation of distortion with output power is shown in Fig. 9.

Intermodulation distortion, measured with carrier and modulating frequencies of 10kc/s and 40c/s respectively, is better than 1·5%.

Frequency Response

The frequency-response characteristic of each channel of the prototype amplifier, measured for an output of 1W is shown in Fig. 8. At 20kc/s, the response is 3dB below the level at 1kc/s and at 15c/s it is 1·5dB down. The response at the rated output power of 7W is 3dB down at 15c/s.

Sensitivity

The sensitivity of each channel of the circuit (including the controls) with 21dB of negative feedback is 100mV for the rated output of 7W.

Hum and Noise

The level of hum and noise in each channel of the prototype amplifier is better than 65dB below the rated output.

Tone Controls

A continuously variable treble-cut tone control is included in each channel of the amplifier. The characteristic of this control is shown in Fig. 8, from which it can be seen that full application of the control gives about 23dB of cut at 10kc/s.

D.C. CONDITIONS

The d.c. voltages in each channel should be tested with reference to Table 1. The results shown in this table were obtained with an Avometer No. 8.

CHAPTER 14
Three-valve Stereophonic Amplifier

The circuit drawn in Fig. 1 has been designed to provide dual-channel stereophonic amplification of a reasonably good quality at a fairly low cost. Only one channel of the amplifier is shown in Fig. 1. The circuitry appearing between the dotted lines is for the left-hand channel only: it should be duplicated for the right-hand channel. The circuitry drawn outside the dotted lines (the power supply, for example) is common to both channels. Each channel uses one Mullard triode pentode, type ECL82, the triode section of which is used for voltage amplification and the pentode section for power amplification. The h.t. supply for both channels is obtained from one Mullard full-wave rectifier, type EZ80.

The input sensitivity of each channel for the rated output of 2W is 280mV, which is adequate for use with the majority of stereophonic crystal pick-up heads. This sensitivity has been achieved in spite of the small complement of valves by using only a small amount (approximately 6dB) of feedback in each channel. The distortion occurring when a continuous sine-wave signal is applied to the input is about 5%, but under the normal input conditions of speech or music signals, the level of distortion will be somewhat lower. The amplifier, although not comparable in specification with high-quality equipment, has been found to give most pleasing stereophonic results during listening tests in normal-sized living rooms. The feedback voltage is taken from the secondary winding of the output transformer of each channel and injected in the cathode circuit of the triode section of the ECL82 in that channel.

CIRCUIT DESCRIPTION

Resistors and capacitors appearing in the left-hand channel of the amplifier are numbered 1, 2, 3, etc.; the corresponding components in the right-hand channel are numbered 101, 102, 103, etc.

Input Selector Switch

The input stages of both channels are connected to the 3-way selector switch SA. The switch positions indicated in Fig. 1 provide the following facilities:

(a) Stereophonic reproduction from stereophonic crystal pick-up heads.

(b) Dual-channel monaural reproduction from a monaural pick-up head. In this position, the left-hand pick-up input terminal is 'live', and both channels are connected in parallel at the input. The input

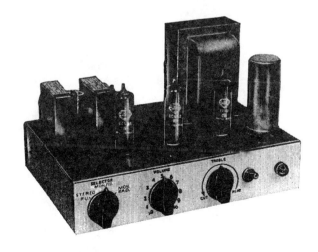

Prototype of Three-valve Stereophonic Amplifier

3-VALVE STEREOPHONIC AMPLIFIER

terminal for the right-hand channel of a stereophonic pick-up head is earthed at position *b* of SA2. If position *b* of SA3 is earthed instead of being connected to position *b* of SA1, single-channel reproduction is possible.

(c) Dual-channel monaural reproduction from an f.m. tuner unit. The input socket in Fig. 1 is connected for monaural applications. If position *c* of SA3 is connected to the right-hand input terminal instead of the left-hand terminal, the circuit will be suitable for reproducing stereophonic transmissions. If position *c* of SA3 is earthed instead of being connected to the input socket, the system gives single-channel monaural reproduction from an f.m. tuner unit.

Input Stage

The triode section of one ECL82 is used in the first stage of each channel for voltage amplification, providing a stage gain of about 50 times. A volume control RV1 (RV101) is included in the grid circuit of the triode section. A simple tone-control network, consisting of the components RV2 and C1 (RV102 and C101) is also incorporated in this grid circuit. The control provides continuously variable treble cut and, with full application of RV2 (RV102), gives at least 20dB cut at 10kc/s. Dual-ganged potentiometers are used in the prototype for the volume-control components RV1 and RV101 and also for the tone-control components RV2 and RV102 so that equal adjustments can be made simultaneously to both channels. It may be found more convenient to use dual-concentric potentiometers for the volume and tone controls because these also allow separate adjustments to be made to the channels.

Output Stage

The output stage of each channel consists of the pentode section of an ECL82. Unless precautions are taken, the acoustical outputs from the loudspeakers of the two channels will not be exactly the same. In order that allowance can be made for any differences, a dual-ganged balance control (RV7 and RV107) is included in the grid circuits of the pentode sections of the ECL82s. This control is made up of a logarithmic-law potentiometer connected in reverse in one channel and an antilogarithmic-law potentiometer connected normally in the other channel. The characteristics of this type of control are discussed on page 28. If dual-concentric potentiometers are used for the volume control RV1, RV101, balance between the channels can be achieved with them, and the potentiometers RV7, RV107 can be replaced by fixed resistors (470kΩ).

Power Supply

A conventional power supply using the Mullard full-wave rectifier, type EZ80, with resistance-capacitance

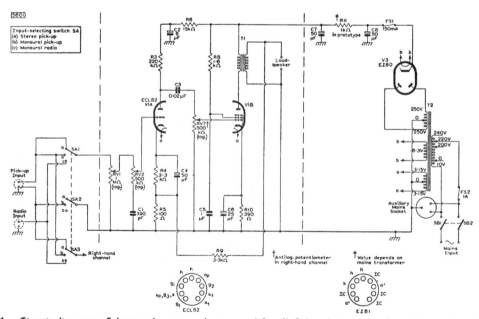

Fig. 1—Circuit diagram of three-valve stereophonic amplifier (left-hand channel only is shown: circuitry between vertical dotted lines should be repeated in the right-hand channel)

3-VALVE STEREOPHONIC AMPLIFIER

smoothing provides the h.t. for both channels. The supplies are taken from the junction of the smoothing components C7 and R11, these components and the reservoir capacitor C8 being common to both channels. The total h.t. current drain of the amplifier is 86mA at 220V and the total heater current is 2·56A at 6·3V.

CONSTRUCTION AND ASSEMBLY

The chassis for the 2W stereophonic amplifier is made from one piece of 16s.w.g. aluminium sheet 14 in. long and 11 in. wide. This piece should be marked as shown in the drawings of Fig. 2, and the holes should be cut as indicated.

Most of the smaller components are mounted on two tagboards. The components for the right-hand channel are arranged on Tagboard No. 1, and those for the left-hand channel are mounted on Tagboard No. 2. Details of these boards are given in Figs. 4 and 5 respectively. It will be observed that the arrangement of the components on one board is the 'mirror-image' of the arrangement on the other.

The larger components such as switches, valve-

LIST OF COMPONENTS

(Resistors and capacitors in the left-hand channel are numbered 1, 2, 3, etc.; corresponding components in the right-hand channel are numbered 101, 102, 103, etc.)

Resistors

Circuit ref.		Value		Tolerance (±%)	Rating (W)
[1]RV1 and	RV101	2× 1	MΩ	logarithmic potentiometer	
[1]RV2 and	RV102	2×500	kΩ	logarithmic potentiometer	
R3,	R103	220	kΩ	10	½
R4,	R104	3·3	kΩ	10	½
R5,	R105	100	Ω	10	½
R6,	R106	15	kΩ	10	½
[1]RV7,		500	kΩ	logarithmic potentiometer	
[1]RV107		500	kΩ	antilogarithmic potentiometer	
R8,	R108	1·8	kΩ	10	½
R9,	R109	3·3	kΩ	10	½
R10,	R110	390	Ω	10	1
[2]R11		1	kΩ	10	6

[1] 10% law (See text)
[2] Value depends on choice of Mains Transformer

Valves

Mullard ECL82 (two), EZ80

Valveholders

B9A (noval) (three). McMurdo BM9/U

Mains Transformer

Primary: 10–0–200–220–240V
Secondaries: H.T. 250–0–250V, 80mA
L.T. 3·15–0–3·15V, 2A (for ECL82)
0–6·3V, 1A (for EZ80)

Commercial Components

Manufacturer	Type No.
Colne	04023
Elden	1299
Gardners	RS.3103
Gilson	W.O.1289
Hinchley	1449
Howells	MM22
Parmeko	P2930
Partridge	P4134
Savage	5D49
Wynall	7920

Capacitors

Circuit ref.		Value		Description	Rating (V)
C1,	C101	390	pF	silvered mica	
C2,	C102	8	μF	electrolytic	350
C3,	C103	0·02	μF	paper	250
C4,	C104	50	μF	electrolytic	6
C5,	C105	1	μF	paper	250
C6,	C106	25	μF	electrolytic	25
C7		50	μF	electrolytic	350
C8		50	μF	electrolytic	350

Tolerance of silvered-mica capacitor is 10%

Miscellaneous

Mains input plug, 3-pin. Bulgin, P429
Mains switch, 2-pole. N.S.F., 8370/B3
Fused voltage selector. Clix, VSP/393/2, P62/1
Auxiliary mains socket, 3-pin. Bulgin, P438
H.T. supply plug (pre-amplifier) 6-pin. Bulgin, P427
Fuseholder. Belling Lee, L.356
Fuse, 150mA
Lampholder, miniature. Bulgin, D675/1/Red
Pilot lamp, 6·3V, 40mA
Input socket, 2-pin Screenector (two). Belling Lee, L.789/CS
Output socket, 2-pin, (two). Painton, 313263
Tagboard, 10-way (two). Bulgin, C125, Denco
Selector switch, 3-pole, 3-way, miniature rotary

Output Transformer (two)

Primary: 5kΩ
Secondary: 3·75, 7·5 and 15Ω

Commercial Components

Manufacturer	Type No.
Colne	03091
Elden	1300
Gardners	AS.7003
Gilson	W.O.767
Hinchley	1534
Howells	MO22
Parmeko	P2928
Partridge	P4135
Wynall	7921

3-VALVE STEREOPHONIC AMPLIFIER

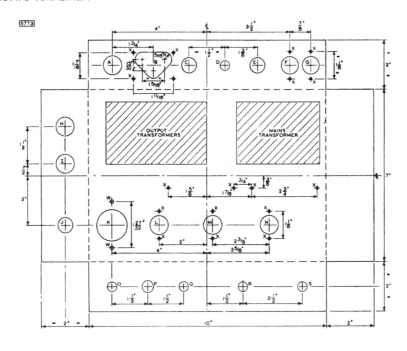

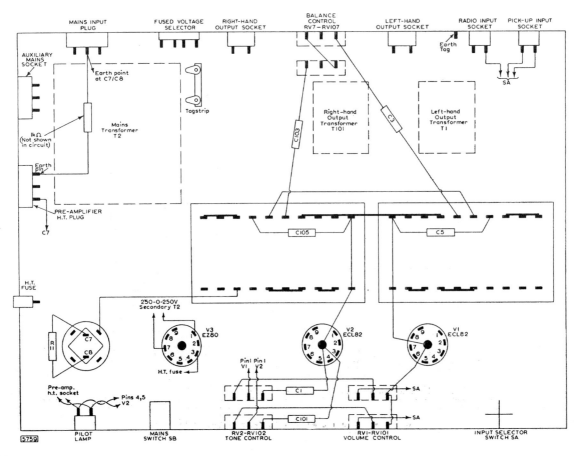

3-VALVE STEREOPHONIC AMPLIFIER

KEY TO HOLES IN CHASSIS

Hole	Dimension	Use	Type No.
A	¾ in. dia.	Mains input plug, 3-pin. Bulgin	P429
B	—	Voltage selector, fused. Clix	VSP/393/2
C	⅜ in. dia.	Output socket, 2-pin. Painton	313263
D	⅜ in. dia.	500kΩ log./antilog. potentiometers	—
E	⅜ in. dia.	Output socket, 2-pin. Painton	313263
F	11/16 in. dia.	Input socket, 2-pin Screenector. Belling Lee	L.789/CS
G	11/16 in. dia.	Input socket, 2-pin Screenector. Belling Lee	L.789/CS
H	¾ in. dia.	Auxiliary mains socket, 3-pin. Bulgin	P438
I	¾ in. dia.	H.T. supply plug, 6-pin. Bulgin	P427
J	½ in. dia.	Fuseholder. Belling Lee	L.356
K	1¼ in. dia.	Electrolytic capacitor	—
L	¾ in. dia.	B9A valveholder. McMurdo	BM9/U
M	¾ in. dia.	B9A valveholder. McMurdo	BM9/U
N	¾ in. dia.	B9A valveholder. McMurdo	BM9/U
O	⅜ in. dia.	Lampholder. Bulgin	D675/1/Red
P	1½ in. dia.	Mains switch. N.S.F.	8370/B3
Q	⅜ in. dia.	2×1MΩ logarithmic potentiometers	—
R	⅜ in. dia.	2×500kΩ logarithmic potentiometers	—
S	⅜ in. dia.	Selector switch, 3-way, miniature rotary	—
W	1½ in. dia.	—	—
X	2·9 mm dia.	—	—

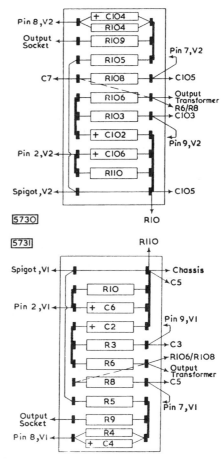

Fig. 2 (top left)—Chassis details (chassis should be bent up at 90° at all dotted lines)

Fig. 3 (bottom left)— Suggested layout of components

Fig. 4 (top right)—Tagboard No. 1 (components of right-hand channel)

Fig. 5 (right)—Tagboard No. 2 (components of left-hand channel)

(Below) Underside View of Prototype Amplifier

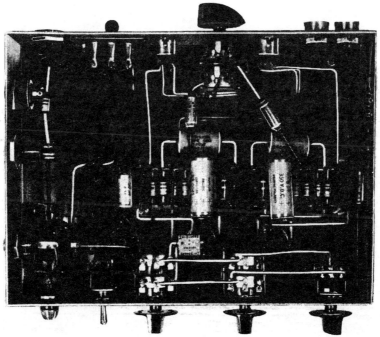

3-VALVE STEREOPHONIC AMPLIFIER

holders and transformers should be mounted on the chassis in the position indicated in the layout diagram of Fig. 3. The recommended wiring between these components is also shown in Fig. 3. The wires connecting the input sockets to the selector switch should be taken along the edge of the chassis beneath Tagboard No. 2. This arrangement follows the practice normally adopted in audio equipment of separating by as much as possible the sensitive input circuit from the rest of the amplifier.

PERFORMANCE
Distortion
The total harmonic distortion was measured in the prototype amplifier with a continuous sine-wave input signal at 400c/s. With 6dB of negative feedback the distortion for the rated output of 2W per channel is approximately 5%. The level of distortion for speech or music input signals is somewhat lower than this. The variation of distortion with output power is plotted in Fig. 6.

Frequency Response
The frequency-response characteristic of each channel of the prototype amplifier is shown in Fig. 7. This response is 3dB below the response level at 1kc/s at frequencies of 40c/s and 40kc/s.

Sensitivity
The sensitivity of each channel of the circuit with 6dB of negative feedback is 280mV for the rated output of 2W.

Hum and Noise
The level of hum and noise in each channel of the prototype amplifier is at least 62dB below the rated output.

Tone Control
The tone-control characteristic is given in Fig. 8. From this it can be seen that 23dB of treble cut are available at 10kc/s with full application of the control.

D.C. CONDITIONS
The d.c. voltages in each channel should be tested with reference to Table 1. The results shown in this table were obtained with an Avometer No. 8.

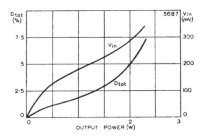

Fig. 6—Variations of harmonic distortion, D_{tot} and input voltage V_{in} with output power

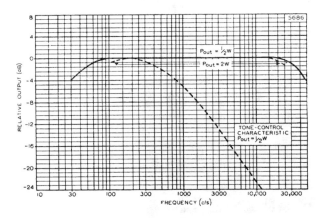

Fig. 7—Frequency-response, power-response and tone-control characteristics

TABLE 1

D.C. Conditions in Each Channel

Point of Measurement		Voltages (V)	D.C. Range of Avometer* (V)
Common to both channels	C8	293	1000
	C7	223	1000
	C2	217	1000
ECL82	Pentode anode	213	1000
	Pentode screen grid	213	1000
	Pentode cathode	13	100
	Triode anode	120	1000
	Triode cathode	1·5	25

*Resistance of Avometer:
1000V-range, resistance = 20MΩ
100V-range, resistance = 2MΩ
25V-range, resistance = 500kΩ

CHAPTER 15

Stereophonic Pre-amplifier

The circuit diagram for a high-quality dual-channel stereophonic pre-amplifier is drawn in Fig. 1. Only one channel is shown in the figure. The circuitry which appears between the dotted lines is for the left-hand channel and it should be duplicated for the right-hand channel. The switches and sockets which are drawn outside the dotted lines are common to both channels.

The circuit is fundamentally a combination of two two-valve pre-amplifiers (Chapter 9), so that the total complement of valves is four low-noise pentodes, type EF86. Each channel is arranged to give an output voltage high enough to drive an amplifier built to the 20W design. A simple potential divider (Fig. 2) can be used to attenuate this output to the level required for amplifiers built to either the 10W circuit or the 3W circuit. An auxiliary output taken from a tapping point on the anode load of the second EF86 is available in each channel for programme monitoring.

Facilities have been provided in this equipment for magnetic and crystal pick-up heads, tape-recorder playback heads and radio tuner units. An auxiliary socket for any input source convenient to the user is also provided in the equipment.

The halves of the input sockets are connected to the switch SB which selects one input at a time. This switch also short-circuits the unused sockets to earth, an arrangement which reduces considerably the amount of 'break-through' between inputs. The positions of the switch, from left to right, are: Auxiliary, Radio, Tape, Microgroove and 78 r.p.m.

The equalisation for disc recordings conforms to the present R.I.A.A. characteristics which have been adopted by most major recording companies. The tape playback characteristic is intended for use with high-impedance heads when replaying pre-recorded tapes at a speed of $7\frac{1}{2}$ inches per second. The tone controls used in each channel cover a wide range of frequency and provide boost and cut for high and low frequencies. The switch SA permits either or both channels of the pre-amplifier to be used. In its central position, both channels are available. In position a, the switch short-circuits the left-hand channel to earth while in position c, the right-hand channel is inoperative. The switch SC between the volume control RV28 (RV128) and the output to the power amplifiers connects the two channels for normal or reversed* stereophonic reproduction or for dual-channel monaural reproduction.

---

*Reversed stereophony is achieved when the left- and right-hand input signals are conveyed to the right- and left-hand loudspeakers respectively.

Prototype of Stereophonic Pre-amplifier

STEREOPHONIC PRE-AMPLIFIER

CIRCUIT DESCRIPTION

Each channel of the pre-amplifier is made up of two stages, and each stage uses a Mullard high-gain pentode, type EF86. All the equalisation takes place in the first stage, and is achieved by means of frequency-selective feedback between the anode and the grid of the first EF86. There is no feedback in the second stage, and the output from the anode of the second EF86 is taken by way of the capacitor C12 to a passive tone-control network.

This arrangement has been chosen because the grid-circuit impedance of the first stage should be low. A low impedance at this grid lessens hum pick-up and reduces the effect of plugging-in external low-impedance circuits. Furthermore, the arrangement also results in low gain in the first stage. Hence, Miller effect between the anode and grid of the first EF86, which can be troublesome when high values of resistance are used in series with the grid, is reduced.

Series resistors are used in the input circuit so that the sensitivity and impedance of any input can be adjusted accurately. The component values given in Fig. 1 are intended for sources encountered most frequently, but the sensitivity and impedance* for each input can be altered by changing the value of the appropriate series resistor.

Maximum output is obtainable with the arrangement shown in Fig. 1, and the output voltage developed is sufficient to drive the 20W circuit described in Chapter 5. This output voltage can be attenuated to the level required to drive the 10W or 3W circuits of Chapters 6 or 7 by introducing the appropriate network shown in Fig. 2 between the volume control RV28 and the switch SC in Fig. 1. An auxiliary output signal, which is suitable for monitoring programmes, is taken from a point on the anode load of the second EF86.

To compensate for any differences in acoustical output that may occur in the two amplifying channels, a balance control consisting of RV16 and RV116 is included between the two stages. This consists of a logarithmic potentiometer connected in reverse in one channel ganged to an antilogarithmic potentiometer connected normally in the other channel. The characteristics of such a control are discussed on page 28.

Dual-ganged potentiometers are used in the prototype for the tone and volume controls. It may be found more convenient to use dual-concentric potentiometers which allow separate adjustments to be

---

*The impedance at each input includes the grid impedance of the EF86 modified by the feedback components as well as the impedance of the input network.

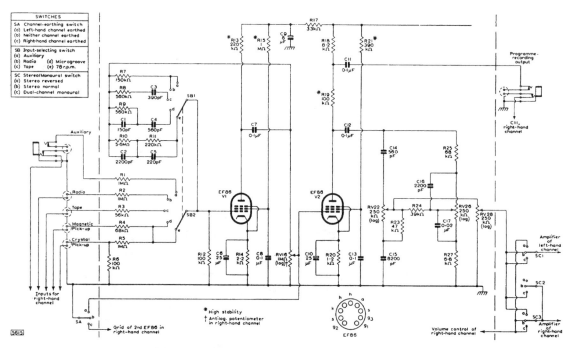

Fig. 1—Circuit diagram of stereophonic pre-amplifier (left-hand channel only is shown: circuitry between vertical dotted lines should be repeated in right-hand channel)

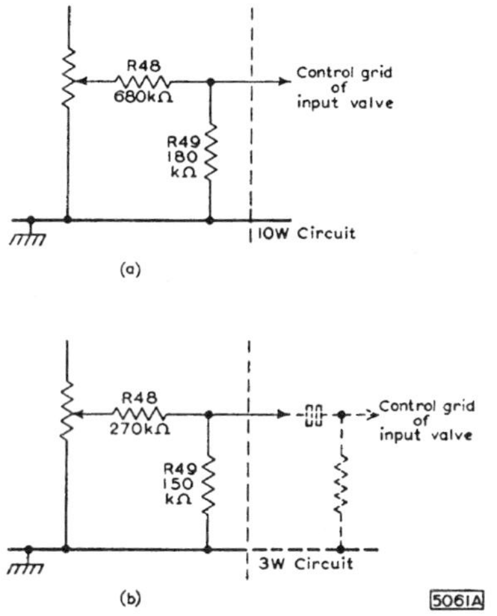

Fig. 2 (left)—Output-attenuating network

TABLE 1

H.T. Smoothing Components for Each Channel

Power Amplifier	Smoothing Resistor (kΩ±10%, ½W)	Decoupling Capacitor (μF)
20-watt	56	16
10-watt	22	16
3-watt	22	16

LIST OF COMPONENTS

Resistors and capacitors in the left-hand channel are numbered 1, 2, 3, etc.; corresponding components in the right-hand channel are numbered 101,102, 103, etc.

Resistors

Circuit ref.		Value	Tolerance (±%)	Rating (W)
R1,	R101	1 MΩ	10	¼
R2,	R102	1 MΩ	10	¼
R3,	R103	56 kΩ	10	¼
R4,	R104	68 kΩ	10	¼
R5,	R105	1 MΩ	10	¼
R6,	R106	100 kΩ	10	¼
R7,	R107	150 kΩ	10	¼
R8,	R108	560 kΩ	10	¼
R9,	R109	560 kΩ	10	¼
R10,	R110	5·6MΩ	10	¼
R11,	R111	220 kΩ	10	¼
R12,	R112	100 kΩ	10	¼
[1]R13,	R113	220 kΩ	10	½
R14,	R114	2·2kΩ	10	½
[1]R15,	R115	1 MΩ	10	½
[2]RV16		1 MΩ logarithmic potentiometer		
[2]RV116		1 MΩ antilogarithmic potentiometer		
R17,	R117	33 kΩ	10	½
[1]R18,	R118	8·2kΩ	10	½
[1]R19,	R119	100 kΩ	10	½
R20,	R120	1·2kΩ	10	½
[1]R21,	R121	390 kΩ	10	½
[2]RV22 and	RV122	2×250 kΩ logarithmic potentiometer		
R23,	R123	47 kΩ	10	¼
R24,	R124	39 kΩ	10	¼
R25,	R125	68 kΩ	10	¼
[2]RV26 and	RV126	2×250 kΩ logarithmic potentiometers		
R27,	R127	6·8kΩ	10	¼
[2]RV28 and	RV128	2×250 kΩ logarithmic potentiometers		

[1]High stability, cracked carbon
[2]10% law (See text)

Capacitors

Circuit ref.		Value	Description	Rating (V)
C1,	C101	150 pF	silvered mica	
C2,	C102	2200 pF	silvered mica	
C3,	C103	390 pF	silvered mica	
C4,	C104	560 pF	silvered mica	
C5,	C105	220 pF	silvered mica	
C6,	C106	25 μF	electrolytic	12
C7,	C107	0·1 μF	paper	350
C8,	C108	0·1 μF	paper	350
C9,	C109	8 μF	electrolytic	350
C10,	C110	25 μF	electrolytic	12
C11,	C111	0·1 μF	paper	350
C12,	C112	0·1 μF	paper	350
C13,	C113	0·1 μF	paper	350
C14,	C114	560 pF	silvered mica	
C15,	C115	8200 pF	silvered mica	
C16,	C116	2200 pF	silvered mica	
C17,	C117	0·02μF	paper	350

Tolerance of all silvered mica capacitors is ±10%

Valves

Mullard EF86 (four)

Miscellaneous

H.T. supply plug, 6-pin. Bulgin, P427
Input socket, Screenector (five). Belling Lee, L.789/CS
Output socket, Screenector (two). Belling Lee, L.789/CS
Jack socket, 3-pole (two). Bulgin, S14
Channel-earthing switch, 1-pole, 3-way
Specialist Switches, SS/593/A; A B Metals, L02
Input-selecting switch, 3-pole, 5-way
Specialist Switches, SS/593/B
Stereo/Monaural switch, 1-pole, 3-way
Specialist Switches, SS/593/C; A B Metals, L02
Set of three switches, Shirley Laboratories, SBL S/S/196
(Note: Details of proprietary switches may not be identical with those given in diagrams.)
Tagboard (10-way) (four). Bulgin, C125; Denco
Tagboard (5-way) (two). Bulgin, C120; Denco

Valveholders

B9A nylon-loaded, with screening skirt. McMurdo, XM9/AU, skirt 95 (four)

made to the channels. If this type is used for the volume control RV28, RV128, there will be no need for the balance control RV16, RV116, which can thus be replaced by fixed resistors (1MΩ).

The d.c. supply of the power amplifiers can be used to provide the h.t. supply for the channels of the pre-amplifier. The smoothing components should normally be mounted in the chassis of the power amplifier used with each channel. The values of these components will vary with the type of power amplifier used, and suitable values are indicated in Table 1. The h.t. current drawn by each channel of the pre-amplifier is 3mA at 230V, and the l.t. current is 0·4A at a heater voltage of 6·3V.

CONSTRUCTIONAL DETAILS

The chassis and layout of the pre-amplifier have been designed specifically for the home constructor. A conventional box-type chassis is not used. Instead, the chassis is made on the unit system, the separate parts being joined together during the assembly of the equipment.

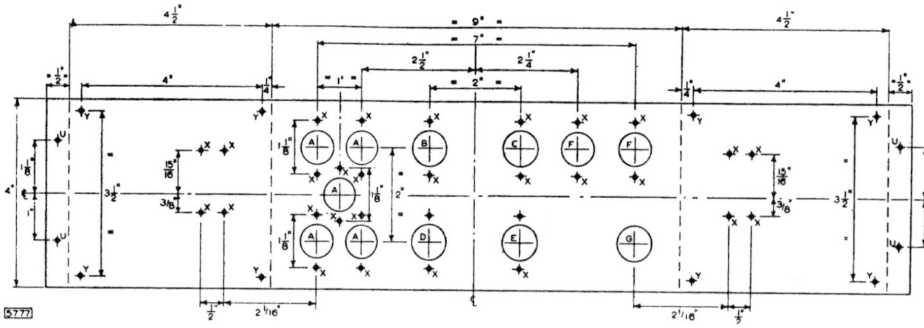

(a)

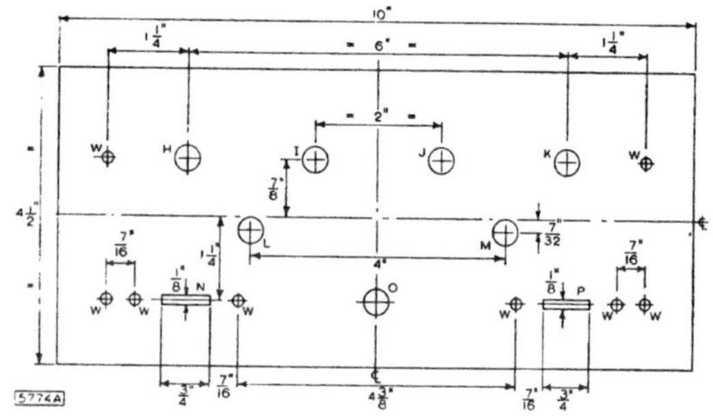

(b)

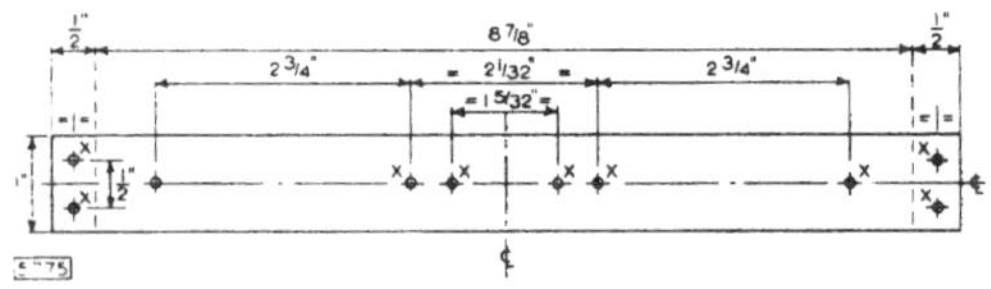

(c)

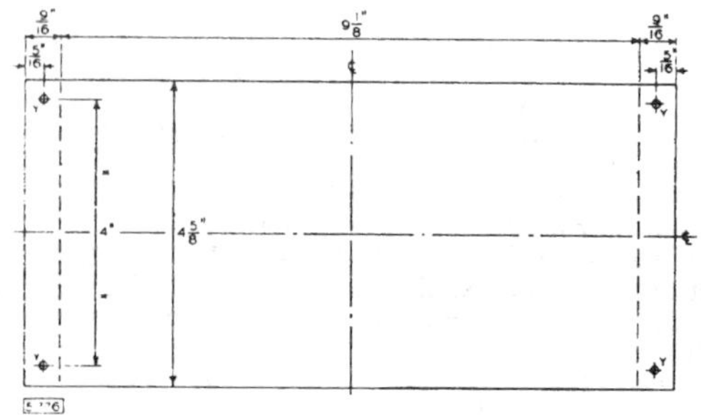

(d)

Fig. 3—Chassis details (the pieces should be bent up at 90° at all dotted lines)

(a) Rear

(b) Front

(c) Tagboard mounting strip (two)

(d) Cover (two)

KEY TO HOLES IN CHASSIS

Hole	Dimension	Use	Type No.
A	11/16 in. dia.	Input socket, Screenector. Belling Lee	L.789/CS
B	3/4 in. dia.	B9A (noval) nylon-loaded valveholder with screening skirt. McMurdo ..	XM9/AU Skirt 95
C	3/4 in. dia.	B9A (noval) nylon-loaded valveholder with screening skirt. McMurdo ..	XM9/AU Skirt 95
D	3/4 in. dia.	B9A (noval) nylon-loaded valveholder with screening skirt. McMurdo ..	XM9/AU Skirt 95
E	3/4 in. dia.	B9A (noval) nylon-loaded valveholder with screening skirt. McMurdo ..	XM9/AU Skirt 95
F	11/16 in. dia.	Output socket, Screenector. Belling Lee	L.789/CS
G	3/4 in. dia.	H.T. supply plug, 6-pin. Bulgin	P427
H	3/8 in. dia.	Selector switch	—
I	3/8 in. dia.	2×250kΩ logarithmic potentiometers	—
J	3/8 in. dia.	2×250kΩ logarithmic potentiometers	—
K	3/8 in. dia.	2×250kΩ logarithmic potentiometers	—
L	3/8 in. dia.	Input jack. Igranic	S14
M	3/8 in. dia.	Output jack. Igranic	S14
N	3/4 × 1/8 in.	Channel-earthing switch	—
O	3/8 in. dia.	250kΩ log./antilog. potentiometer ..	—
P	3/4 × 1/8 in.	Stereo/monaural switch	—
U	6·7 mm dia.	4BA hank-bushes	—
X	2·9 mm dia.	—	—
Y	Drill No. 49	—	—

STEREOPHONIC PRE-AMPLIFIER

The chassis is made up of six separate pieces of 16 s.w.g. aluminium sheet, the dimensions (in inches) of which are as follows:

- (a) Rear panel 19×4
- (b) Front panel $10 \times 4\frac{1}{2}$
- (c) Mounting strip (two) $9\frac{7}{8} \times 1$
- (d) Cover plate (two) $10\frac{1}{4} \times 4\frac{5}{8}$

Each piece should be marked as shown in Fig. 3 and the holes should be cut as indicated. It is important that, when bending the pieces, the scribed lines should lie exactly along the angles. This ensures that the pieces will fit together properly when assembled.

For ease of assembly, components should be mounted on tagboards, and these should then be bolted to the mounting strips before the strips are attached to the rear plate. Diagrams showing the positions of the components on the tagboards are given in Figs. 4, 5, 6 and 8, 9, 10.

When the mounting strips and tagboards have been attached to the rear plate, connections should be made between the valveholders and the components on the

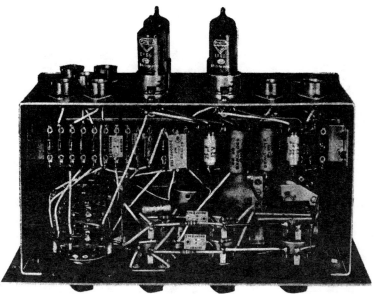

Top View of Prototype Pre-amplifier

Underside View of Prototype Pre-amplifier

STEREOPHONIC PRE-AMPLIFIER

tagboards. (The valveholders should be mounted so that the valves will be on the outside of the completed chassis, and the valves should be fitted with screening cans.)

The ganged potentiometers RV22-RV122, RV26-RV126 and RV28-RV128 should be mounted on the front panel of the chassis, and the other components which make up the tone-control network should be connected to them. The balance control RV16-RV116 and the switches SA, SB, SC should also be connected to the front panel, which should then be bolted to the back panel. The remaining components should be connected in position as indicated in the general layout diagrams of Figs. 7 and 11. Details of the selector switch are given in Fig. 12.

PERFORMANCE

The values for hum and noise in the pre-amplifier which are quoted for each input position have been measured with each channel connected to a 20W power amplifier. The measurements were made at the output socket of the power amplifier when the input terminals of the pre-amplifier were open-circuited. The frequency-response curves were also obtained with

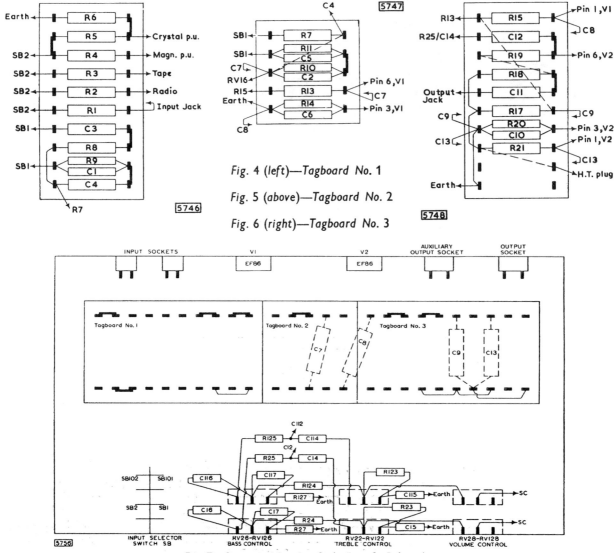

Fig. 4 (left)—Tagboard No. 1

Fig. 5 (above)—Tagboard No. 2

Fig. 6 (right)—Tagboard No. 3

Fig. 7—Suggested layout of components—top view

STEREOPHONIC PRE-AMPLIFIER

this combination of pre-amplifier and power amplifier.

The sensitivity figures given below provide an output from the pre-amplifier of 250mV when the full anode load of the second EF86 is used. All measurements were made with the balance control set for balance.

PICK-UP INPUT POSITIONS

Equalisation curves for magnetic and the crystal pick-up positions are drawn in Fig. 13. The difference in sensitivities between the positions for microgroove and 78 r.p.m. records is achieved by the different amounts of feedback provided at positions d and e of the switch SB1 (SB101).

Magnetic Pick-up Position

Input impedance	100kΩ (approx.)
Sensitivity at 1kc/s	
(a) microgroove	5mV
(b) 78 r.p.m.	15mV
Hum and noise	
(a) microgroove	58dB below 20W
(b) 78 r.p.m.	59dB below 20W

This arrangement is most suitable for pick-up heads

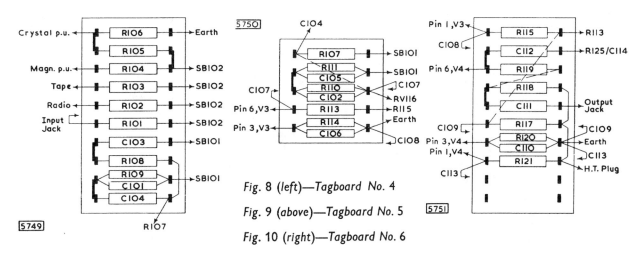

Fig. 8 (left)—Tagboard No. 4

Fig. 9 (above)—Tagboard No. 5

Fig. 10 (right)—Tagboard No. 6

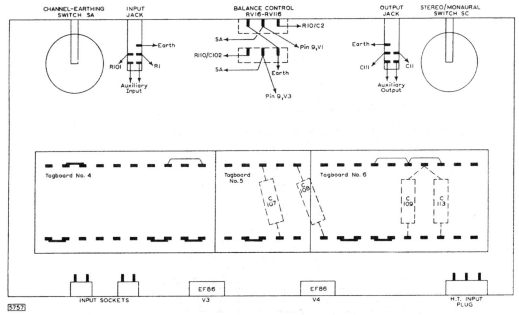

Fig. 11—Suggested layout of components—underside view

STEREOPHONIC PRE-AMPLIFIER

of the variable-reluctance type, but moving-coil types which have higher outputs can be used if a larger value of series resistance R4 is included.

Crystal Pick-up Position

Input Impedance	100kΩ
Sensitivity at 1kc/s	
(a) microgroove	70mV
(b) 78 r.p.m.	210mV
Hum and Noise	
(a) microgroove	58dB below 20W
(b) 78 r.p.m.	59dB below 20W

Low- and medium-output crystal pick-up heads can be used with this position. The input is loaded with the 100kΩ resistor R6 in order that its characteristic shall approximate to that of a magnetic cartridge, and to allow the same feedback network to be used. This produces the best compromise with most types of pick-up head. However, if the head is not suitable for this form of loading, or if its output is too high, then it can be connected to the auxiliary input socket, the function of which is discussed below.

TAPE PLAYBACK POSITION

Input Impedance	80kΩ (approx.)
Sensitivity at 5kc/s	4·0mV
Hum and Noise	54dB below 20W

The equalisation characteristic used for this position is shown in Fig. 14. For frequencies above 100c/s, the curve follows the C.C.I.R. characteristic, but below

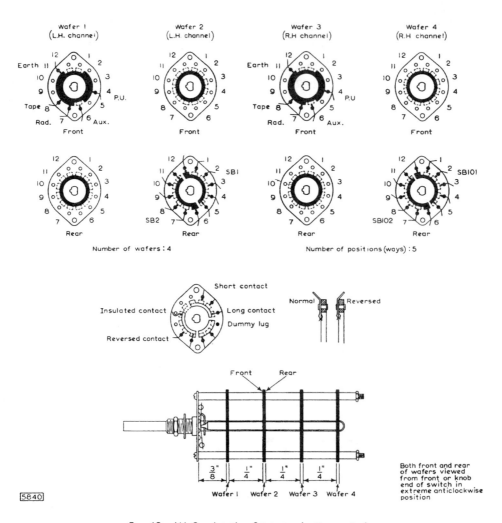

Fig. 12—Wafer details of input-selecting switch

STEREOPHONIC PRE-AMPLIFIER

this frequency, slightly less boost is used. The channel is intended for replaying pre-recorded tapes using high-impedance heads, and the characteristic adopted results in good performance with these heads. If a greater sensitivity is required, the value of the resistor R3 can be decreased until the desired sensitivity is obtained.

RADIO INPUT POSITION

Input Impedance	1MΩ
Sensitivity	330mV
Hum and Noise	58dB below 20W

The frequency-response characteristic for this position is given in Fig. 17. With the values of impedance and sensitivity quoted above, this channel should meet most requirements. Other values can easily be obtained, however, by altering the feedback resistor R7 and the series resistor R2. If the input impedance of the channel is too high, it can be reduced by connecting a resistor of the appropriate value between the input end of R2 and the chassis.

AUXILIARY INPUT POSITION

It can be seen from the circuit of Fig. 1 that the auxiliary position is identical with the radio input channel. The input with the component values shown in Fig. 1 can therefore be used for high-output crystal pick-ups, for example, or for tape pre-amplifiers such as the circuit described in Chapter 12. The auxiliary input is taken to a jack socket on the front panel of the equipment. The insertion of the jack disconnects the coaxial input socket on the back of the chassis.

TONE CONTROLS

The treble and bass tone-control characteristics of the pre-amplifier are shown in Fig. 16. These indicate that an adequate measure of control is provided in the unit for most applications.

Low-impedance controls have been adopted so that any capacitance resulting from the use of long coaxial leads between the pre-amplifier and main amplifier will have a minimum effect on the output impedance of the pre-amplifier.

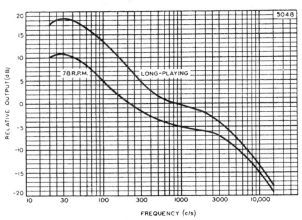

Fig. 13—*Equalisation characteristic for pick-up input signals*

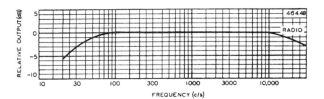

Fig. 15—*Frequency-response characteristic for radio input signals*

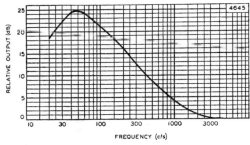

Fig. 14—*Equalisation characteristic for tape playback input signals*

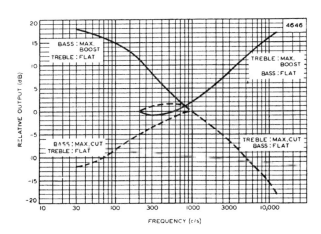

Fig. 16—*Tone-control characteristics*

STEREOPHONIC PRE-AMPLIFIER

HARMONIC DISTORTION

The total harmonic distortion of each channel of the pre-amplifier is less than 0·15% at normal output levels. At outputs of ten times this level, the harmonic distortion is only 0·24%.

AUXILIARY OUTPUT POSITION

An additional output from the second EF86 of each channel is available at this position, enabling a record of programme material to be made with tape equipment. This additional output is taken to a jack socket at the front of the chassis. The voltage at the auxiliary output socket of each channel is about 250mV and this output is available at a low impedance. Recording equipment plugged into this socket should not have an impedance less than 500kΩ. The tone controls are inoperative when this output is used.

D.C. CONDITIONS

The d.c. voltages in each channel should be tested with reference to Table 2. The results shown in this table were obtained using an Avometer No. 8.

TABLE 2

D.C. Conditions in Each Channel

Point of Measurement		Voltage (V)	D.C. Range of Avometer* (V)
H.T.		230	1000
C9		200	1000
2nd EF86	Anode	68	1000
	Screen grid	118	1000
	Cathode	2·2	25
1st EF86	Anode	60	1000
	Screen grid	70	1000
	Cathode	1·7	25

*Resistance of Avometer:
1000V-range, resistance = 20MΩ
25V-range, resistance = 500kΩ